Characterization of Progesterone and its Receptors in Ovarian Tumors "Benign and Malignant"

Prof Dr. Sami AL-Mudhaffar
Dr. Saba Z. AL- Qadi

Summary

1. The level of progesterone was determined by radioimmunoassay (RIA) in sera of normal women and patients with benign and malignant serous ovarian tumors.

2. The presence of progesterone receptors was investigated in two groups of serous ovarian tumors (benign and malignant). The obtained results indicate a higher incidence of these receptors in both benign and malignant tumors.

3. A radio receptor assay for crude progesterone receptors was developed using ^{125}I-progesterone and found to be suitable for the assessment of those receptors in serous ovarian tumors.

4. The characteristics of the binding of ^{125}I-progesterone with its receptors in benign and malignant serous ovarian tumors homogenate were investigated. Different factors affecting this binding were extensively studied such as the concentration of receptors, the concentration of ^{125}I-progesterone, pH, temperature, time, halides and salts.

5. The time-course of ^{125}I-progesterone binding with its receptors in benign and malignant serous ovarian tumors homogenate was studied at four different temperatures. The obtained data have obeyed pseudo-first order kinetics for the association of ^{125}I-progesterone with its receptors.

6. The kinetic parameters K_{+1}, K_{-1}, Ka and Kd of the binding of ^{125}I-progesterone with its receptors in benign and malignant serous ovarian tumors homogenate were determined at four different temperatures.

7. Thermodynamic parameters of the standard and transition states of ^{125}I-progesterone binding with its receptors in benign and malignant serous ovarian tumors homogenate were determined. These parameters have been used to characterized the binding reaction.

8. The progesterone receptors were isolated from benign and malignant serous ovarian tumors homogenate by gel filtration technique.

9. Spectroscopic studies in the U.V. range (190-350 nm) were carried out on isolated progesterone receptors from benign and malignant serous ovarian tumors, and the effect of pH and polarity were also studied.

List of Contents

Chapter One: Introduction and Literature survey

Chapter Two: Experimental Work

Chapter Three: Results and Discussion

List of Abbreviations

-A-

A----------------------------Absorbance
AR--------------------------Androgen receptor

-B-

B----------------------------Bound
B_{max}--------------------------Maximal binding capacity
BSA-------------------------Bovine serum albumin

-C-

CA_{125}---------------------Carbohydrate antigen-125
CBG--------------------- Corticosteroid-binding globulin
CEA-------------------------Carcinoembryonic antigen
CPM-------------------------Count per minute

-D-

DCC---------------------------Dextran-coated charcoal
DMSO----------------------Dimethyl sulfoxide

-E-

ER----------------------------Estrogen receptor

-F-

F----------------------------Free
FSH--------------------------Follicle stimulating hormone

-H-

H----------------------------Hormone
hCG-------------------------Human chorionic gonadotrophin
hrs---------------------------Hours

-K-

Ka----------------------------Affinity constant
Kd----------------------------Equilirium dissociation constant
KD---------------------------Kilo dalton

-L-

LDL----------------------Low density lipoprotein
LH-----------------------Luteinizing hormone

-M-

MSTI--------------------- Molar surface tension increment

-N-

NSB----------------------Non-specific binding

-p-

PEG---------------------Poly ethylene glycol
PR----------------------Progesterone receptor

-R-

R-------------------------Receptor
RIA----------------------Radioimmunoassay
RRA---------------------Radioreceptorassay
RSB---------------------Relative specific binding

-S-

SB----------------------Specific binding

-T-

TB----------------------Total binding
TC----------------------Total count

-U-

U.V.--------------------Ultra Violet

List of Figures

List of Tables

CHAPTER ONE
INTRODUCTION
&
LITERATURE SURVEY

Introduction

1.1 Progesterone

Progesterone is a C_{21} steroid (figure 1-1), produced in the ovaries and in the placenta [1], it controls the growth, development and physiology of the female reproductive tract and other organ systems.

It was responsible for induction of secretory activity and decidual development in the endometrium of the estrogen-primed uterus and it's required for the implantation of the fertilized ovum and maintenance of pregnancy [2]. In non-pregnant women, progesterone secretion is largely confined to cells of the corpus luteum, but because it is an intermediate in biosynthesis of all steroid hormones, small amount may also be released from the adrenal cortex. Some progesterone is also produced by granulosa cells just before ovulation [3,4]. The first 6-8 weeks of the pregnancy, the corpus luteum is the major source of progesterone then the placenta assumes this function [5].

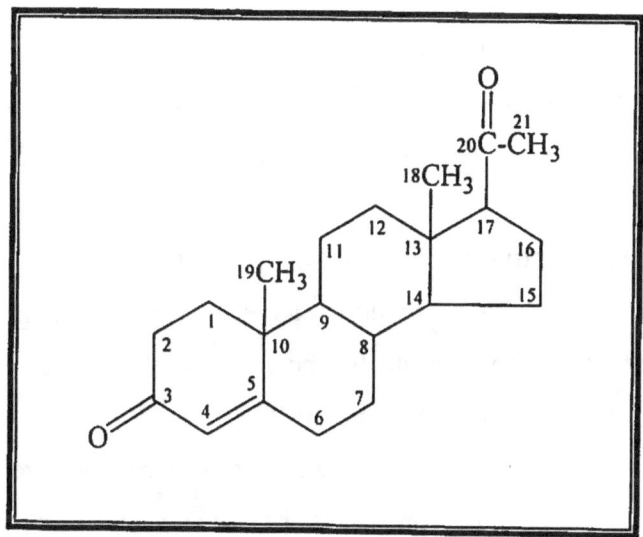

Figure (1-1): Progesterone structure [1].

1.1.1 Chemistry

The structure of progesterone, a C_{21} compound, is shown in figure (1-1). Progesterone (pregn-4-ene-3,20-dione) contains a keto group at (C-3) and double bound between (C-4 and C-5) both characteristics are essential for progestational activity. The side chain (CH_3CO-) on (C-17) dose not seem to be very important for its physiological action [6].

1.1.2 Biochemistry and physiology

1.1.2.1 Biosynthesis

Progesterone is synthesized from cholesterol, which is present in the gland both free and esterified to fatty acids (cholesteryl esters). Cholesterol which is derived either from circulating lipoproteins (LDL) or from cholesteryl esters, in the gland it is converted to pregnenolone by the mitochondrial enzyme P_{450scc}, which removes a six-carbon fragment, isocaproic acid.

The pregnenolone formed by this reaction may be converted either to progesterone or to 17α-hydroxy-pregnenolone. The conversion to progesterone requires the action of 3β –hydroxy steroid dehydrogenase in which dehydrogenation of the OH radical to the ketone radical and $\triangle^{5,4}$-ketosteroid isomerase shifts the double bound from the $\triangle^5$ to the $\triangle^4$ position [7,8].

The synthesis of progesterone is shown in figure (1-2) [9].

In luteal tissue, low-density lipoprotein cholesterol is thought to serve as the preferred precursor despite the potential of the corpus luteum to synthesize progesterone de novo from acetate [10].

Initiation and control of luteal secretion of progesterone are regulated by LH and FSH [11,12], LH increases the conversion of cholesterol to progesterone [13].

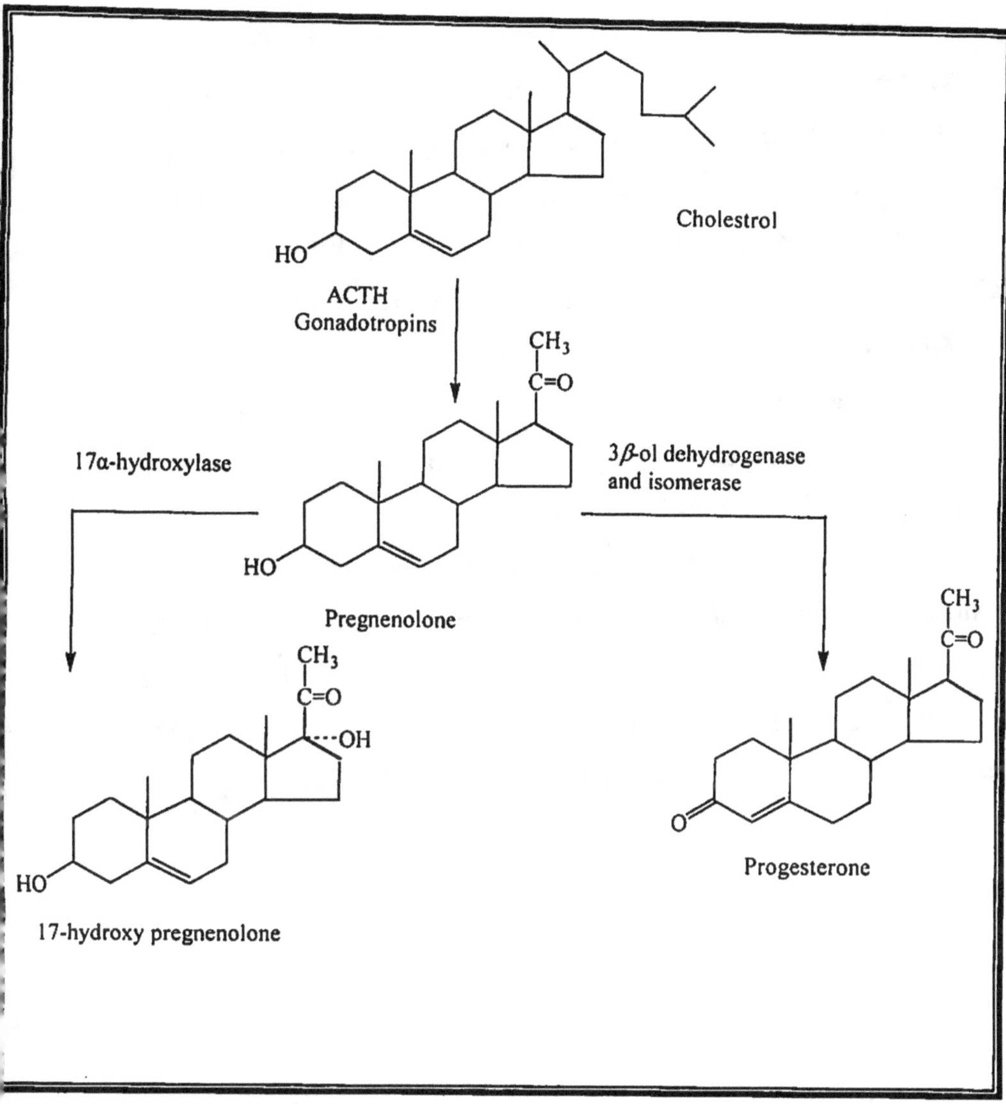

Cholestrol

ACTH
Gonadotropins

CH_3
$C=O$

17α-hydroxylase

3β-ol dehydrogenase
and isomerase

HO

Pregnenolone

CH_3
$C=O$
----OH

HO

17-hydroxy pregnenolone

CH_3
$C=O$

O

Progesterone

Figure (1-2): Pathway of progesterone biosynthesis [9].

1.1.2.2 Transport

Progesterone, like all steroids hormones, is water insoluble compound that is transported in the blood stream to its side of action by association with plasma carrier proteins [14,15]. It doesn't have a specific plasma protein but it is, like cortisol, bound to corticosteroid-binding globulin [16] and it is bound strongly to corticosteroid-binding globulin (CBG) and weakly to albumin [7].

1.1.2.3 Metabolism

The important metabolic event leading to inactivation of progesterone is reduction and conjugation [6]. Progesterone is rapidly cleared from the circulation having an initial half-life of about 5 minutes.

It is rapidly converted to pregnanediol and conjugated to glucuronic acid in the liver. Pregnanediol glucuronide is excreted in the urine and may be used an index of progesterone production. In addition, small amount of 20α-hydroxy progesterone is formed, this compound has one-fifth the activity of progesterone [7]. Metabolism of progesterone is shown in figure (1-3).

Figure (1-3): Metabolism of progesterone [6,7]

1.1.3 Mechanism of progesterone action

The principle target organs of progesterone are the uterus, the breast and the brain [17]. The hormone is bound to its receptors within target cells. The mechanism of action, figure (1-4), is as follows [18-20]:

1. A lipid soluble hormone diffuses from the blood, through interstitial fluid, and through the lipid bilayer of the plasma membrane into a cell.

2. If the cell is a target cell, the hormone will bind to activate receptors located within the cytosol or nucleus. An activated receptor then alters gene expression: It turns specific genes of the nuclear DNA on or off.

3. As the DNA is transcribed, new messenger RNA (mRNA) forms, leaves the nucleus, and enter the cytosol where it directs synthesis of new proteins, usually enzymes, on the ribosomes.

4. The new proteins alter the cells activity and cause the typical physiological responses of the hormone.

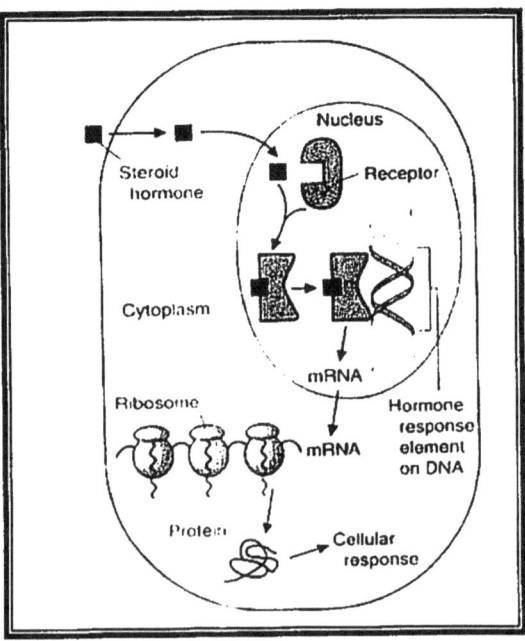

Figure (1-4): Mechanism of action of hormones that bind to intracellular receptors [21].

1.1.4 Progesterone function

Progesterone is the most important hormone associated with pregnancy, it's responsible for both the successful initiation and the successful completion of pregnancy. It prepares the lining of the uterus to accept the fertilized egg. Once the egg is attached, progesterone is involved in the development of the fetus and plays a role in the suppression of further ovulation during pregnancy, and progesterone also inhibits uterine contraction, increases the viscosity of cervical mucus, promotes glandular development of the breast and increases basal body temperature (BBT) [1,22,23]. It also causes retention of water, sodium and nitrogen [24].

1.1.5 Levels of progesterone during normal menstrual cycle

Progesterone was secreted as part of the normal menstrual cycle [25], the term menstrual cycle refers to the series of changes that occur in sexually mature, nonpregnant females and culminate in menses. Typically, the menstrual cycle is approximately 28 days long, although it may be as short as 18 days in some women and as long as 40 days in others [26].

The menstrual cycle is usually divided into two phases [27]:

1. Follicular phases or "Proliferative phase".

2. Luteal phase or "Secretory phase".

- **Follicular phase:** The levels of progesterone is approximately (0.9 ng/ml) (3 nmole/ml) during this phase. Late in this phase, after ovulation which is occurs around day 14 of the menstrual cycle, progesterone secretion begins to increase and continues to increase [26].

  In vitro work has shown that the production of progesterone is correlated with the hormonal environment during the follicular phase of the cycle [28].

- **Luteal phase:** During this phase, the corpus luteam produces large quantities of progesterone and ovarian secretion increases about 20 fold. The result is an increase in plasma progesterone to a peak value of approximately (18 ng/ml) (60 nmole/ml) [29].

  The stimulating effect of LH on progesterone secretion by the corpus luteum is due to activation of adenylate cyclase and involves a subsequent step that is dependent on protein synthesis [30].

  The hormone levels during the menstrual cycle is shown in figure (1-5).

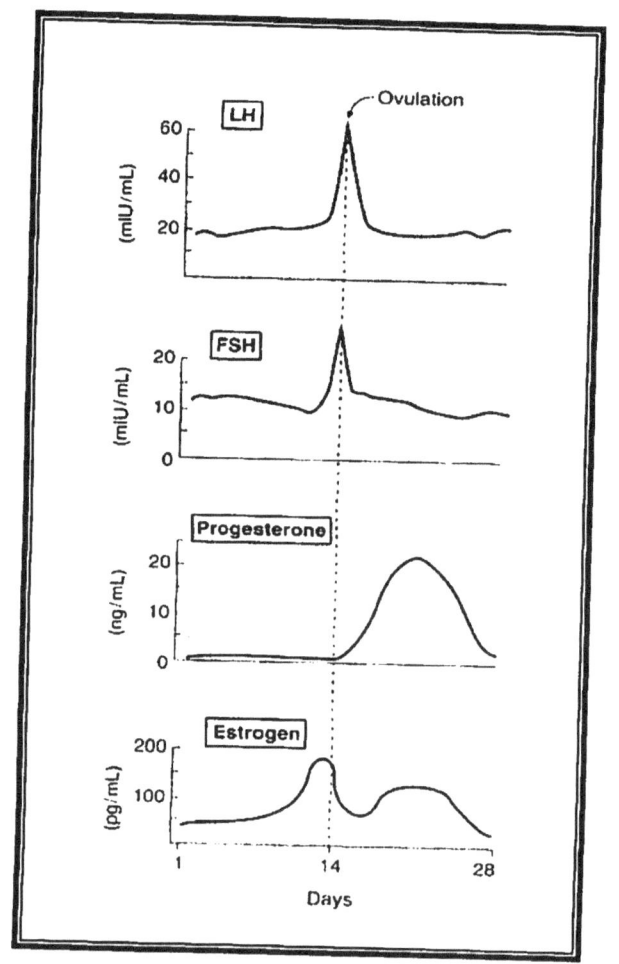

Figure (1-5): Hormone levels during the menstrual cycle [21].

1.1.6 The role of progesterone in the hypothalamic-pituitary-ovarian axis.

Progesterone and estrogens are important in the control of the hypothalamic-pituitary-ovarian axis, figure (1-6) [31]. The relationship among the hypothalamus, pituitary and ovary in the control of gonadal function mean abnormality in any of these organs may cause abnormal menstruation and infertility. The release of LH and FSH is affected both positively and negatively by estrogen and progesterone; whether estrogen and progesterone stimulate or inhibit gonadotropin release depends on concentration and duration of exposure of the pituitary to steroids [32,33].

The gonadotrophin secretion regulated by the ovarian hormones feed beck at the level of pituitary [34].

- **Negative feed back:** ovarian steroid and peptide hormones can exert a negative feed back on both the hypothalamus and pituitary [35,36].

The increase in LH and FSH occurs when ovarian secretion of estrogen decreases after menopause or castration [37] but the inhibition of FSH and LH secretion occurs at low levels of estrogen. Progesterone at high concentrations inhibits FSH and LH primarily at the level of the hypothalamus [38].

- **Positive feed back:** ovarian steroid and peptide hormones are also able to exert positive feed back which is important in the regulation of the LH surge required to induce ovulation and is regulated by sharply rising levels of estrogen in the late first half of the menstrual cycle [39].

Progesterone is reportedly for the mid cycle FSH surge [40] and low concentrations stimulates LH release, but only after previous prolonged exposure of the pituitary to estrogen [41].

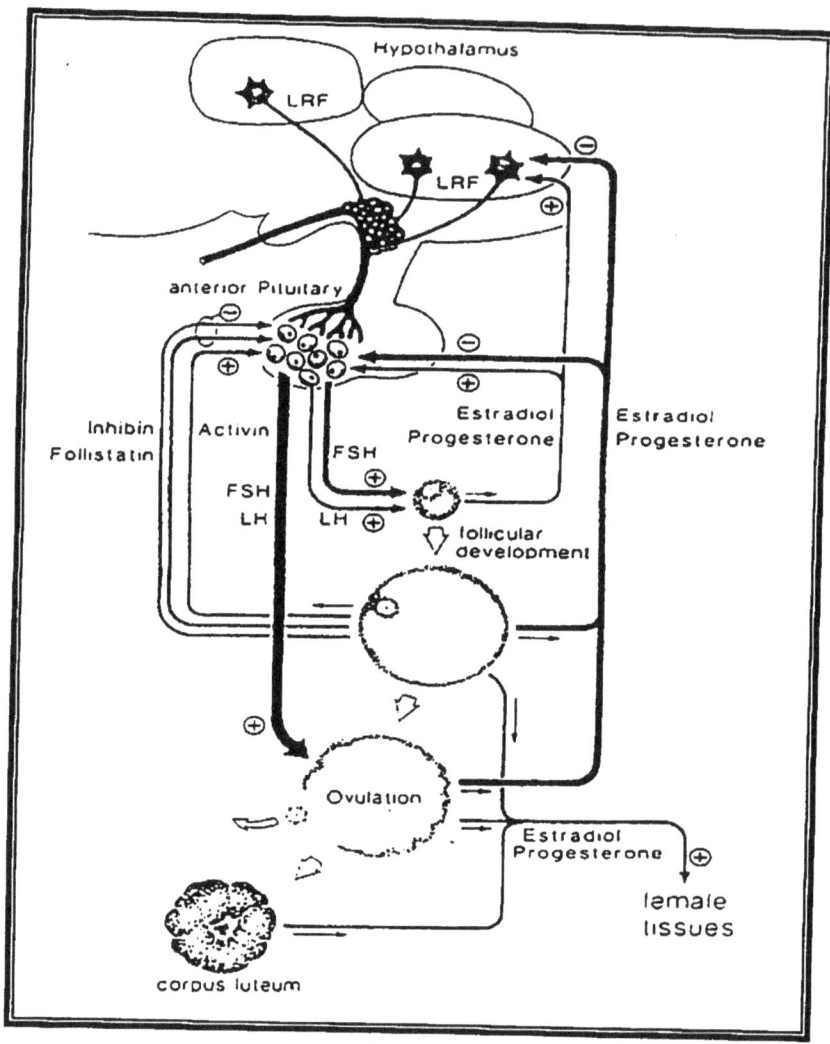

Figure (1-6): A diagrammatic representation of the hypothalamic-pituitary-ovarian axis[31].

1.1.7 Progesterone receptors

Progesterone receptors are specific proteins located in the nucleus of the cells which form the target organs for progesterone action, their function is to receive active hormone (progesterone) entering into the cells from the circulation and subsequently mediated the intracellular response to the hormone [42].

The progesterone receptor is an acidic protein with an isoelectric point around pH 5 [43]. It has a molecular weight of about 110,000 daltons [44].

Several methods are available to detection of progesterone receptors (PR); most of these methods employ the same basic principles which include the incubation of tumor homogenate with radioactive hormone and determination of the amount of radioactivity bound to receptor protein [45].

However variety of methods are used to separate bound and free hormones [46]:

1. Addition of charcoal suspension to adsorb the unreacted free hormone.

2. Addition of hydroxy-apatite suspension to adsorb the bound hormone.

3. Precipitation of bound hormone by the addition of protamine sulphate.

4. Gel filtration.

5. Electrophoresis.

6. Sucrose density gradient analysis.

Receptor assays in biopsis removed from cancer patients have demonstrated the presence of hormone receptors in almost (60-70%) of tumor-sensitive cells and their strongly indicates that these patients have a favorable prognosis and a better response to hormone therapy [47]. Conversely, patients with hormone negative receptors do not respond to hormone therapy and have a poor prognosis with greater propensity to metastasis [48,49]. Therefore, estrogen receptor (ER) and progesterone receptor (PR)-binding proteins have been demonstrated to be necessary for steroid hormonal function in steroid target and their malignancies and study of cytoplasmic steroid receptor content allows insight into the cellular regulation of malignancies [50].

Several studies have been demonstrated the presence and importance of progesterone receptors in different human organs like breast [51] uterus [52,53] and ovary [54-56].

The human progesterone receptor level was found to be a prognostic indicator in ovarian cancer, independent of patients age, stage of disease, histological type and grade of differentiation [57].

1.2 Ovarian tumors

Ovarian tumors are common, and they are benign but most malignant, ovarian tumors are the leading cause of death from reproductive tract cancer [58].

The tumors may be solid, cystic or a mixture of both, and may be benign, malignant or in a borderline state [59]. According to the bases of distinct clinical and pathologic features, ovarian tumors can be divided into three major categories [60,61] (table 1-1)[62]:

- Epithelial tumors.
- Germ cell tumors.
- Sex cord-stromal tumors.

About 80% of ovarian tumors are benign, and these occur mostly in young women between the ages of 20 and 45 years. The malignant tumors are more common in older women, between the ages of 40 and 65 years [62].

The vast majority of ovarian tumors are epithelial tumors, these account 90% of malignant ovarian tumors [60,63].

Table (1-1): Ovarian tumors [World Health Organization (WHO) classification][62].

Surface epithelial-stromal tumors	Serous tumors	Benign (cystadenoma) Of borderline malignancy Malignant(serous cystadenocarcinoma) Cystadenocarcinoma
	Mucinous tumors, Endocervical-type and intestinal-type	Benign (of borderline) malignancy Malignant
	Endometrioid tumors	Benign Of borderline malignancy Malignant Epithelial-stromal Adenosarcoma Mesodermal (mullerian) Mixed tumor
	Clear cell tumors	Benign Of borderline malignancy Malignant
	Transitional cell tumors	Brenner tumor Brenner tumor of borderline Malignancy Malignant brenner tumor Transitional cell carcinoma (non-Brenner type)
	Undifferentiated carcinoma	
Sex cord-stromal tumors	Granulosa-stromal cell tumors	Granulosa cell tumors Tumors of the thecoma-fibroma group
	Sertoli-stromal cell tumors (androblastomas)	
	Sex-cord tumor with annular tubules	
	Gynandrobladtoma	
	Steroid (lipid) cell tumors	
Germ cell tumors	Dysgerminoma	
	Yolk sac tumor (endodermal sinus tumor)	
	Mixed germ cell tumors	
Malignant (NOS)		
Metastatic nonvarian cancer (from nonovarian primary)		

NOS = Not otherwise specified.

1.2.1 Some types of ovarian tumors

- **Epithelial ovarian tumors**

Epithelial tumors are the most common type of ovarian neoplasms, the designation of epithelial tumors is from the origin of these tumors-cells derived from the coelomic epithelium of the ovary, the surface cells and their adjacent stroma undergo neoplastic transformation [64]. The classification of common epithelial tumors has been developed by Word Health Organization (WHO) and the International Federation of Gynecology Obstetrics (FIGO) [60,65]. According to the (WHO), surface epithelial-stromal tumors can be divided into: serous tumors, mucinous tumors, endometrioid tumors, clear cells tumors, transitional cell tumors and undifferentiated carcinoma [66,67], and epithelial tumors have been subdivided into benign, borderline and malignant [68-70].

Epithelial tumors of the ovary metastasize primarity by direct extension, implantation of tumor cells on the peritoneal surface and lymphatic spread [71].

1.2.2 Incidence

Epithelial ovarian cancer (EOC) accounts for (80-90%) of all ovarian malignances [72], so it is the fifth most common cause of death from cancer in women [73-75].

The incidence of ovarian cancer increases with age, being greatest between ages 65 and 84.

Approximately 70% of cases will occur in women who are over the age of 50 year [76,77]. According to the data of Iraqi Cancer Registry Center [78] collected for the period between (1986-2000), figure (1-7), there is a tendency toward an increase in the frequency of ovarian cancer incidence during the last year, therefore, ovarian cancer is the seventh most common malignant tumors among the women in Iraq.

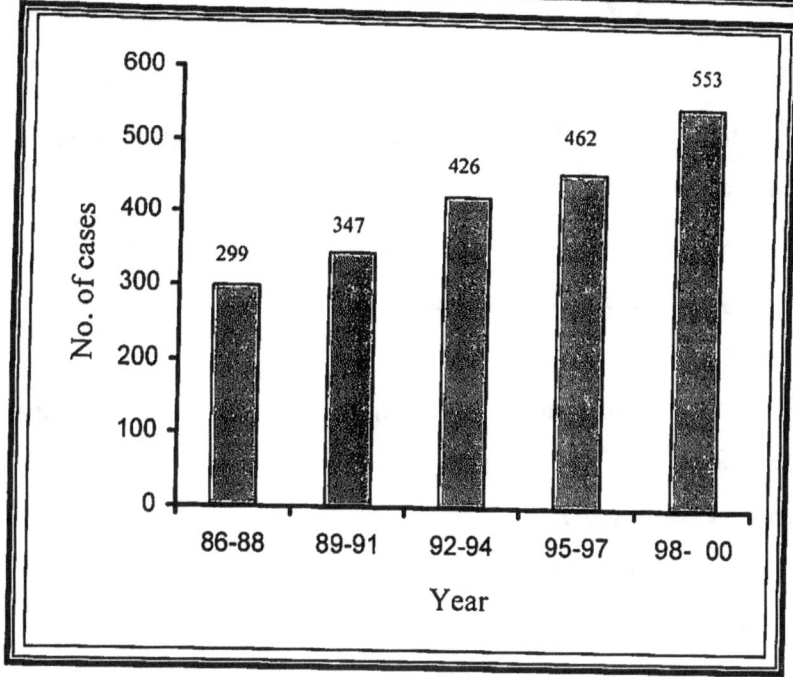

Figure (1-7): **The population of ovarian cancer in Iraq through (1986-2000)** [78].

1.2.3 Etioligy

The etiology of ovarian cancer which is identified as playing an important role in the development of ovarian cancer are: Reproductive factor, Genetic [79-81] and Environmental factors [82,83].

1. Reproductive risk factor:

Associated with an elevated risk for the development of ovarian cancer include those associated with an increased number of ovulation in a woman's lifetime [84,85]. Among women with ovarian cancer, an increased incidence of nulliparity and a lower mean number of pregnancies were detected. In women with one or two children, the risk of developing ovarian cancer is reduced by about 50%, as compared to nulliparous women [86-88]. Conversely a decreased incidence of ovarian cancer after the use of oral contraceptive [89,90]. Women who used oral contraceptive have 40% lower chance of developing ovarian cancer, compared to those who never used them.

2. Genetic:

Approximately (5-10%) of all ovarian cancers are thought to be hereditary in etiology [91], ovarian cancer is a component of three hereditary cancer syndromes [92,93]:

- hereditary breast and ovarian cancer syndrome
- hereditary site-specific ovarian cancer
- the lynch II syndrome

The majority of hereditary cancer results from inherited mutation in two genes BRCA1 and BRCA2 [94,95].

3. Environmental factor:

That has been consistently associated with an increased risk for the disease is the exposure to talc powder on the perineal area [84].

1.2.4 Serous ovarian tumors

Serous epithelial ovarian tumors are the most common form of epithelial neoplasms and a count for 46% of all epithelial tumors [84,96]. Serous cystadenoma are the most common benign ovarian neoplasms, occur most frequently in women (30-50) years of age [97,98], while serous cystadenocarcinoma are the most common malignant ovarian neoplasms [99].

- **Serous cystadenoma**

A cystadenofibroma is usually defined as a cystadenoma in which at least one fourth of the tumor was consists of fibrous stroma. These tumor vary considerably in size but may reach 20 cm in diameter. They are bilateral in 15% of cases [100].

- **Serous cystadenocarcinoma**

It is the commonest primary carcinoma, according to 60% of all cases, and in over half the cases it is bilateral. The cysts are always of papillary type and the epithelium burrowing through the capsule produces papillary processes on the serous surface. Extension of growth to the pelvis and adjacent organs fixed the tumor. Ascites is always present [101].

1.2.5 Staging

The staging of ovarian tumor is based on extent of spread of tumor and histologic evaluation of the tumor [102]. The International Federation of Gynecology and Obstetrics (FIGO) classification of ovarian cancer is presented in table (1-2) [60,102,103].

Table (1-2): Staging of ovarian cancer according to (FIGO) [103].

Stage	Description
Stage I	**Growth limited to the ovaries**
IA	Growth limited to one ovary, no ascites; no tumor on the external surface: Capsule intact
IB	Growth limited to both ovaries, no ascites; no tumor on the external surface: Capsule intact
IC	Tumor stage IA or IB but with tumor on the surface of one or both ovaries, with capsule ruptured, with ascites present containing malignant cells, with positive peritoneal washing.
Stage II	**Growth involving one or both ovaries with pelvic extension**
IIA	Extension or metastases to the uterus or tubes
IIB	Growth involving one or both ovaries with pelvic extension
IIC	Tumor either stage IIA or IIB but with tumor on the surface of one or both ovaries, with capsules ruptured, with ascites present containing malignant cells, or with positive peritoneal washing.
Stage III	**Tumor involving one or both ovaries with peritoneal implants outside the pelvis or positive retroperitoneal or inguinal nodes, superficial liver metastases equal stage III, tumor limited to the true pelvis but with histologically verified malignant extension to small bowel or omentum.**
IIIA	Tumor grossly limited to the true pelvis with negative nodes but with histologically confirmed microscopic seeding of abdominal peritoneal surface.
IIIB	Tumor of one or both ovaries with histologically confirmed implants of abdominal peritoneal surface, none exceeding 2 cm in diameter; nodes negative.
IIIC	Abdominal implants greater than 2 cm in diameter, or positive retroperitoneal or inguinal nodes.
Stage IV	**Growth involving one or both ovaries with distant metastasis; if pleural effusion is present, there must be positive cytologic test results to allot a case to stage IV.** **Parenchymal liver metastasis equates IV.**

1.2.6 Treatment

The treatment depends upon its pattern of spread

- **Surgical treatment**

Surgery is the primary therapy in the management of patients with ovarian cancers. It is performed initially for the determination of the extant of cancer spread and for the reduction of tumor load [104,105]. Second-look procedures are performed to determine the status of the disease and the response to therapy and to remove detectable remaining tumor masses [106,107]

1. **For benign tumors:** tumor removal or unilateral oophorectomy is usually performed.

2. **For malignant tumors:** in an early stage the standard therapy is complete surgical staging followed by abdominal hysterectomy and bilateral salping-oophorectomy [108] with omentectomy and selective lymphadenectomy, with more advanced disease, aggressive removal of all visible tumor improves survival [58].

- **Radiotherapy**

Radiotherapy has gradually assumed a more important part in the treatment of gynaecological cancer [109]. Radiotherapy has been used as a primary treatment for non-resectable ovarian cancer, as an adjuvant to surgery, for palliation in advanced cases, and for the treatment of recurrent disease. However, the role of radiotherapy in the treatment of ovarian cancer has been the subject of considerable controversy [110,111]. Since ovarian cancer is considered to be a disease of the whole peritoneal cavity, intraperitoneal administration of chromic phosphate (^{32}P) and colloidal gold (^{198}Au) has been used in patients with early stage disease [112].

- **Chemotherapy**

Ovarian cancer is one of the most responsive human malignancies to cytotoxic chemotherapy, combination therapy is shown to achieve a higher response rare than do treatment with single agent [113]. Several chemotheraputic regimens are effective Cisplatin based combination have been the mainstay of treatment for advanced ovarian cancer [114,115].

Acombination of Paclitaxel and Cisplatin appears to be the most effective regimen, it increases the median survival from 24.4 to 37.5 months (54%), over the usual cure in the treatment of advanced epithelial cancer, and appears to be cost-effective first line treatment for advanced ovarian carcinoma.

A synergestic activity of Paclitaxal combind with Ifosfamide, has been reported in ovarian cell lines [116,117].

- **Hormonal therapy**

The fact that the ovary is not only the main source of estrogens and progesterone but also a target organ for these and other hormones, suggests that epithelial ovarian cancer may be an endocrine-related tumor, therefore, to determine the role of hormones in ovarian cancer include empirical treatment with hormonal agents in patients with advanced disease and studies of steroid receptors in ovarian cancer then can replace the conventional cancer therapies (chemotherapy, surgical therapy and radiotherapy) thus becoming new adjuvant systemic therapy for cancer [47,118].

Hormonal treatment with high doses of progesterone have been found to be helpful in patients with ovarian adenocarcinoma and showed subjective and objective improvements during treatment [119,120]. Progesterone may be used as a tumor marker in "non-endocrine" ovarian tumors [121], and have also been used in combination with other agents: like estrogen [122] and Tamoxifen [123].

Hormonal receptors have been also extensively studied in ovarian cancer, estrogen and progesterone receptors [124] are in approximately 50% of tumors [125] and 98% of ovarian cancer contains androgen receptors [126].

In endometrioid cancer of the ovary, the presence of ER and PR is reported to be a reliable criterion in the selection of patients for progestin treatment [127]. Since endocrine treatment is most frequently used after failure of chemotherapy, it is of importance to know the effect of chemotherapeutic agents on receptors levels [128,129]. If failure of hormonal treatment is due to low concentrations of PR, induction of receptors might increase activity [130]. Tamoxifen may increase the number of PR and decrease the number of ER. From this point of view, alternating treatment with Tamoxifen and progesterone might be useful [130].

Finally, the presence of cytoblasmic steroid receptors in ovarian tumors may explain their reported response to endocrine therapy [131].

1.2.7 Diagnosis

1.2.7.1 Symptoms

Early ovarian cancer has an symptomatic character for this reason, ovarian cancer is frequently called a silent killer [132,133].

When symptoms are present they are at first non-specific and many patients are intially refferred to specialist other than gynecologist, many patients had months of gastrointestinal symptoms including belching, early satiety, abdominal fullness. Most patients with ovarian cancer have loss of appetite, ill health and sometimes weight loss and the majority of patients with serous cystadenocarcinoma have advanced disease at the time of diagnosis [14,99,134,135].

The common symptoms of epithelial cancer include abdominal distention due to ascites, pleural effusions may also be present with ascites. Other symptoms include nausea, dyspepsia, lower abdominal discomfort and abdominal vaginal bleeding [64,84].

As the disease progress, symptoms become more specific and constant, related to pain and pressure caused by enlarging tumor or ascites that is not uncommon [136]. Ascites or large omental metastases can lead to dyspepsia and decreased food intake. The weight loss caused by decreased food intake may not be apparent due to abdominal distension by ascites and tumor volume [137].

1.2.7.2 Tumor markers

Tumor markers are substances of different chemical nature that are either produced by a tumor or released by the host response to a tumor, therefore disappearance of most of the marker should indicate eradication of the tumor, whereas an increase in marker concentration should indicate tumor growth [6,138,139].

Tumor markers are useful in screening determining therapy and in detecting relapse [138].

A clssification of tumor markers is provided in figure (1-8) [140]. Tumor markers can be either a biochemical substances, which is normally found in vivo (e.g. hormones and enzymes), or genetically repressed substance; such as oncofetal antigens [141].

Tumor markers might be useful for diagnosis of ovarian cancer and of monitoring the disease status during and after treatment [142,143].

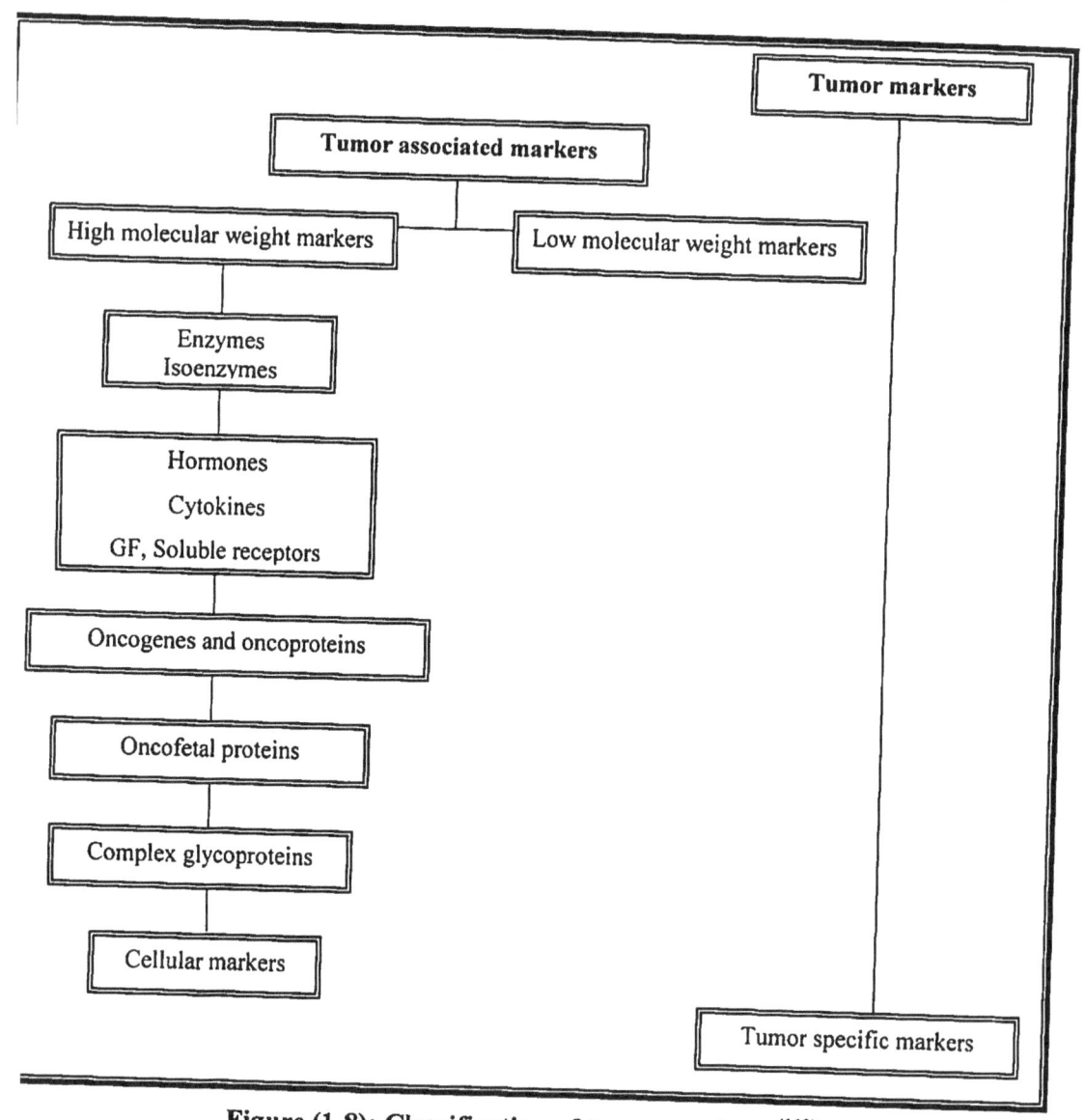

Figure (1-8): Classification of tumor markers [112].

- **CA 125:** Cancer antigen 125 (CA 125) is the most important cancer-associated marker for the management of ovarian cancer and it is the first serum marker test for epithelial cancer of the ovary [96,144]. CA 125 is elevated in 82% of patients with advanced (stage III or IV) ovarian cancer [145,146] and it is also elevated, although less frequently, in women with earlier stage disease [147]. In studies of women with known or suspected ovarian cancer the reported sensitivities of CA 125 in detecting stage I and stage II cancers are 29-75% and 67-100%, respectively [148,149].

So, CA 125 is a useful marker in monitoring the progress of nonmucinous ovarian tumors and is invaluable in the management and follow-up of patients. In follow-up, it reflects the response to therapy (surgery, chemotherapy and radiotherapy) in 80% of cases [150,151].

- **CEA:** Carcenoembryonic antigen (CEA) is a plasma member glycoprotein initially discovered in colonic carcinomas, and the systemic levels of CEA were thought to be specific for this cancer [152]. Elevated serum levels of CEA have also been found in patients with carcinoma of the ovary [153] in particular, mucinous adenocarcinoma [154] and carcinoma of cervix, uterus and breast [155,156]. CEA levels have been useful in the follow-up of ovarian cancer patients [154] to monitor the response of the tumor to chemotherapy [157]. Furthermore, the use of radiolabeled antibodies against CEA can be helpful in the detection of primary and metastatic ovarian tumors [158].

hCG: Human chorionic gonadotrophin (hCG) has been a most effective tumor marker for patients with trophoblastic disease. This hormone consists of alpha and beta subunits, The beta subunit is larger than the alpha subunit and is immunologically distinct [159]. β-hCG is most commonly elevated in normal pregnancy, germ cell tumors and well differentiated endometrial carcinomas [139]. β-hCG were also elevated in selected of epithelial ovarian cancers and immunohistochemical localization of hCG in certain tumors should help to identify those patients in which this marker will be useful clinically [160].

Steroid hormone receptors: Steroid hormone receptors (estrogen, progesterone and androgen receptors) are important determinants of prognosis and predictive behavior in tumor tissues of several origins: breast and endometrial [161,162].

The presence of steroid receptors (ER, PR and AR) has also been correlated with prognosis in common epithelial ovarian cancers [163-165] of the three steroid receptors (ER, PR and AR) only the presence of high levels of PR was found to be associated with prognosis [166].

Several workers found that endometrioid tumors contain more frequently PR, often is association with ER [130,167], while serous tumors were more frequently found to be ER-positive [168] but in muciuons and clear cell tumors [167,169] lower amounts of steroid receptors were detected.

ER-positive tumors are seen more often among postmenopausal women, whereas PR-positive tumors are more frequently seen in premenopausal women [130].

The determination of steroid hormone receptors status offers additional prognostic information in ovarian carcinomas, so that the detection of the hormonal receptors for estrogen (ER) and progesterone (PR) are useful for diagnosing ovarian adenocarcinomas [170,171].

Evaluation of receptor content not only allows a better prediction of prognosis, but it also may be helpful in predicting responses to therapy at the time of a future recurrence.

Aim of the work

The aim of the work in this thesis includes the following:-

1. Determination of progesterone level in sera of normal and patients affected by serous ovarian tumors.

2. Molecular characterization of the binding of ^{125}I-progesterone with its receptors in benign and malignant serous ovarian tumors such as those of binding capacity and the effect of various factors (temperature, time, pH, salts, halides, receptor concentration, hormone concentration).

3. Determination of the kinetic and thermodynamic parameters of the binding reaction of progesterone with its receptors on ovarian tumor in premenopausal patients.

4. Spectroscopic studies on the progesterone receptors in premenopausal patients with benign and malignant serous ovarian tumors.

CHAPTER TWO

EXPERIMENTAL WORK

EXPERIMENTAL WORK

2.1 Chemicals, Instruments and Samples

2.1.1 Chemicals

All common laboratory chemicals and reagents were of analar grade and were used without further purification.

* From BDH

EDTA (disodium salt), (Na,K-tartarate), $CuSO_4.5H_2O$, NaF, $MnCl_2$, Glycerol, Folin cio colteau and β-mercaptoethanol.

* From Fluka

Tris (hydroxymyethyl amino methane), $MgCl_2$, NaCl, Charcoal, Bovine serum albumin and Gelatin.

* From Pharmacia Fine Chemicals, Switzerland

Sephadex G-150, Dextran T-70, Blue dextran 2000.

* From Immunotech (France)

Kit of radioactive progesterone (^{125}I-progesterone). The activity of labeled progesterone was approximately 5 μci.

2.1.2 Instruments

The instruments used in this work were: -

* LKB gamma counter type 1270 Rack gamma II.

* Cooling centrifuge type Hettich.

* Shimadzu U.V-Visible recorder spectrophotometer type U.V-160.

* Pye-Unican pH meter.

* Memmert water bath.

* SM-Shaker.

* Memmert incubator.

2.1.3 Patients

Two groups of epithelial ovarian cancer patients (serous cystadeno carcinoma) and one group of patients with benign epithelial ovarian tumors (serous cystadenoma) were included in this study. **Group I** consist of premenopausal patients with benign ovarian tumors. **Group II** consisted of (8) premenopausal patients with ovarian cancer. **Group III** consisted of (6) postmenopausal patients with the same type of cancer. Two groups of healthy women were used as control.

All patients were admitted for treatment to (Medical City, Baghdad Teaching Hospital), (Al-Kadhimiya Hospital) and Al-Jadirya Hospital. They were histologically proved through the supervision of specialists at Al-Kadhimiya Hospital and Dr. Nawal Alash. The patients were newly diagnosed and not underwent any type of therapy.

Patients suffered from any disease, like hypertension, or diabetes, that may interfere with this study were excluded.

2.1.4 Preparation of blood samples

Blood samples (5-7 ml) were obtained from pre and post-menopausal patients before surgery by veinpuncture. Age matched sera were obtained from healthy female volunteers, (16) premenopausal women and (10) postmenopausal women, were used as control. Blood samples were left for 20 min at room temperature, after coagulation, sera were separated by centrifugation at 1500 xg for 10 min, then sera were aspirated and kept at −20 °C until assaying.

2.1.5 Collection of specimens

The tumor tissues were surgically removed from ovarian tumor patients by hysterectomy. The specimens were cut off and immediately rinsed with ice-cold isotonic saline solution. They were collected individually in plastic receptacles and stored at −20 "C until homogenization.

2.1.6 Preparation of ovarian tumors homogenates

The frozen tissues were weighted, sliced finely with a scalped in petri dish standing on ice bath, and then homogenized in TEMG buffer at pH 7.4 with a ratio of 1:5 (weight-volume) using a manual homogenizer. The homogenate was filtered through four layers of nylon in order to eliminate fibers of connective tissues, then centerifuged at 2000 xg for 30 min at 4 °C. The sediment was suspended in 10 volumes of the TEMG buffer for 15 min at 4 °C and then the suspension was used to obtain the crude nuclear fraction and supernatant was used as crude cytosolic fraction [172].

2.1.7 Buffers and reagents

All buffer solutions were prepared [173] by dissolving the appropriate amount of salt in distilled water and required pH was adjusted.

1. Tris/HCl buffer at different pH values was prepared as followed: -

Solution A: 0.2 M tris (2.4228 g tris [hydroxy methyl] aminomethan) in 100 ml of distilled water.

Solution B: 0.1 N HCl.

Working buffers pH (7.2-9) were prepared by mixing 25 ml of solution A with an appropriate amount of solution B to adjust the required pH, then the volume was made up to 100 ml with distilled water.

2. TEMG buffer (pH 7.4):

0.01 M tris containing 1.5 mM Na_2-EDTA, 2mM of β-mercaptoethanol and 10% glycerol. The buffer was prepared by an appropriate dilution of the stock solution to 500 ml.

3. Dextran-coated charcoal (DCC) suspension:

This suspension was prepared by dissolving the following compound: 1.25 g charcoal, 0.625 g dextran T-70 and 0.2 g gelation in 100 ml of TEMG buffer pH 7.4.

2.2 Determination of progesterone levels in sera of patients with benign and malignant ovarian tumors

Progesterone levels were measured in sera of benign and malignant ovarian tumors patients and healthy individuals were used as controls by radioimmunoassay (RIA). The assay protocol was described in table (2-1).

Table (2-1) : RIA assay protocol of serum progesterone (ng/ml).

Progesterone standard (ng/ml)						Unknown			
0	0.1	0.5	2	1.2	60	(1)	(2)	(3)	(4)
Tube No.									
1,2	3,4	5,6	7,8	9,10	11,12	13,14	15,16	17,18	19,20
Standard serum									
50	50	50	50	50	50	-	-	-	-
Unknown serum									
-	-	-	-	-	-	50	50	50	50
^{125}I-prog.									
500	500	500	500	500	500	500	500	500	500
All volumes are in µl									

All tubes were incubated at room temperature with shaking for 60 min. then the tubes were aspirated and counted by gamma counter for one minute [174].

2.3 Determination of ^{125}I-progesterone concentration

The concentration of labeled progesterone was measured according to the method of Morris [175]: -

In a set of anti-progesterone coated tubes marked from 1 to 12, 500 µl of ^{125}I-progesterone was add with 50 µl of standard unlabeled progesterone of different concentrations (0,0.1,0.5,2,12 and 60 ng/ml). In another set of the same tubes, different volumes (250,350,450,550,650 and 750 µl) of ^{125}I-progesterone were pipetted. After incubation of all tubes for 1 hrs. at 25 °C, they were decanted and counted using gamma counter.

Calculations

1. (B) is the bound radioactivity (cpm) which represents the counted radioactivity in the precipitated hormone-antibody complex.

2. The free hormone (F) which represents unbound ^{125}I-progesterone was determined from the following formula:

   F (cpm) = Total count (cpm)–B (cpm)

3. The values of the ratio (B/F) for an ordinary standard curve were plotted against the concentration of standard progesterone.

4. The (B/F) values for the incubation of different amounts of ^{125}I-progesterone were also calculated, table (2-2).

5. The data in table (2-2) and (2-3) were plotted as in figure (2-1).

6. The amount of radioactivity corresponding to the concentration of unlabeled hormone was estimated by using the two curves (I and II) of figure (2-1). This was carried out by drawing a line which intercepts with both corves at the same increment.

7. The amount of the standard progesterone was plotted against the amount of corresponding radioactivity table (2-4) and the concentration of the ^{125}I-progesterone was determined from the intercept of the straight line as in figure (2-2).

Table   (2-2):-B/F values corresponding to different concentration of
            progesterone standard used in standard curve.

Concentration of standard (ng/ml)	B/F
0	1.53
0.1	1.33
0.5	0.95
2	0.78
12	0.33
60	0.18

Table   (2-3): B/F values corresponding to different amounts of (^{125}I-
            progesterone) used in the incubation.

Amount of ^{125}I-progesterone (cpm) in μl	Bound radioactivity (cpm)	B/F
250	6004	1.42
350	9180	1.08
450	10100	0.962
550	11396	0.81
650	13024	0.63
750	14900	0.47

Table (2-4): The mass of standard progesterone in (ng/ml) corresponding to
            a given amount of the tracer.

Bound radioactivity (cpm)	Hormone concentration (ng/ml)
12240	2
13025	3
13520	4
14000	6
14250	7
14760	8

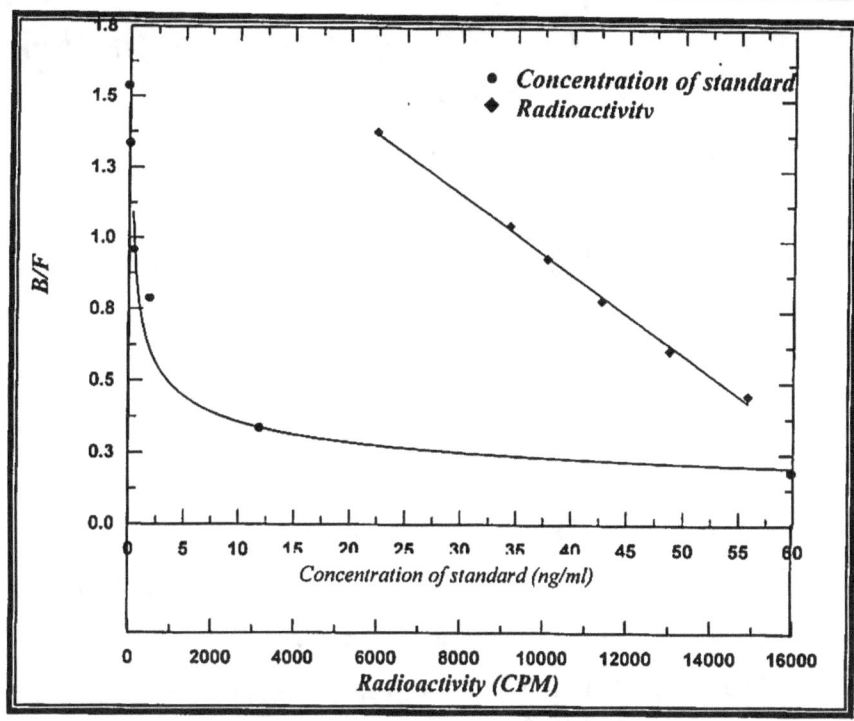

Figure (2-1): Ratio of bound to free radioactivity (B/F) for an ordinary standard curve, where different amounts of standard progesterone were incubated with constant amount of ^{125}I-progesterone and antibody also (B/F) for antibody incubated with different amounts of ^{125}I-progesterone in the absence of unlabeled progesterone.

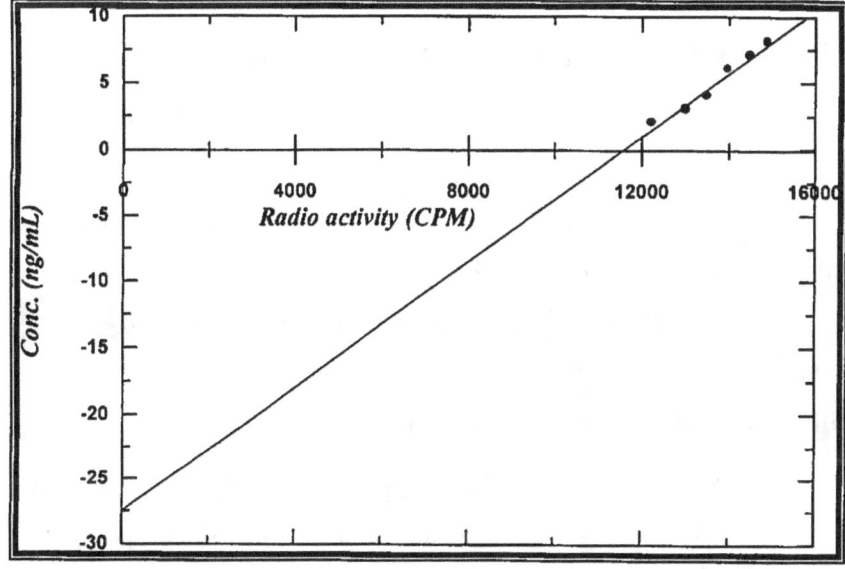

Figure (2-2): A plot of the mass of standard progesterone against cpm of ^{125}I-progesterone having the same B/F values from figure (2-1), which resulted a straight line.

2.4 Binding studies of ^{125}I-progesterone with their receptors in ovarian tumors homogenates.

2.4.1 Estimation of total protein content in tumors homogenates.

Total protein content of ovarian tumors homogenates was determined by the method of lowery et al. [176], using bovine serum albumin (BSA) as the standard protein.

<u>Solutions</u>

1. **Reagent A:** Alkaline sodium carbonate solution (2% Na_2CO_3 in 0.1 N NaOH).

2. **Reagent B:** Copper sulfate-sodium potassium tartarate solution (0.5% $CuSO_4.5H_2O$ in 1% Na,K tartarate). This solution was prepared freshly by dissoliving 0.1 g of Na,K tartarate in 10 ml of $CuSO_4.5H_2O$.

3. **Reagent C:** Alkaline copper solution, this reagent was prepared by mixing 50 ml of reagent A with 1 ml of reagent B.

4. **Reagent D:** Folin Cio Calteau reagent was prepared by the dilution of the commercial reagent with an equal volume of distilled water on the day of use.

5. **Standard bovine serum albumin (stock BSA 0.2 mg/ml):**
   Working BSA solutions were prepared by serial dilution of stock solution.

<u>Procedure</u>

1. One milliliter of each standard bovine serum albumin (20, 40, 60, 80, 100, 120, 200 µg/ml) was pipetted in a set of duplicate tubes.

2. One milliliter of (1:10) diluted tumor homogenate was also pipetted in a duplicate tubes.

3. Five milliliters of reagent C were added to all tubes.

4. The tubes were shacked and allowed to stand at room temperature for 10 min.

5. Half milliliter of reagent D was added to all assay tubes and mixed immediately.

6. The tubes were left at room temperature for 30 min.

7. The absorbance of the blue solution was read at 750 nm against an appropriate blank.

8. The standard curve was obtained by plotting the absorbance against the corresponding concentration of standard protein as shown in figure (2-3), and used in the determination of the unknown protein concentration of the ovarian tumors homogenate.

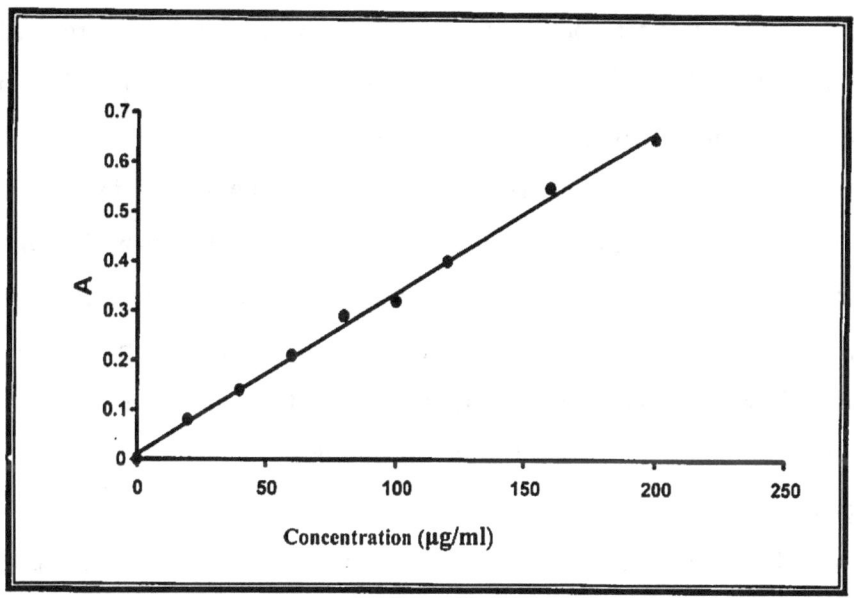

Figure (2- 3): Standard curve of protein determination by lowry's method.

2.4.2 Preliminary test of ^{125}I-progesterone binding to its cytosolic and nuclear receptors in human benign and malignant ovarian tumors.

Cytosolic and nuclear receptors were detected using two sets of experiments. The first one was carried out to determine the total binding, while the second was used for the estimation of non-specific binding. In order to detect cytosolic receptors 200 μl (250 μg protein) of crude cytosol were incubated with 100 μl (6.23 nM) of ^{125}I-progesterone in duplicate tubes. The volumes of the mixture were completed to 1 ml with TEMG buffer pH 7.4 and the tubes were incubated for 16 hrs. at 4°C. Non-specific binding was accounted by preparing the same incubation with the addition of 200 fold excess of unlabeled progesterone as a competitor [177]. After incubation, the bound progesterone was estimated by the dextran-coated charcol method [172]. For this purpose 250 µl of dextran-coated charcol (DCC) were added. The tubes were shaken for 10 min and centrifuge for 10 min at 2000 xg. An aliquot of 1 ml was taken from each supernatant and counted by γ-counter.

Crude nuclear receptors were detected by addition of 200 µl (250 µg protein) of crude nuclear fraction to 100 µl (6.23 nM) of ^{125}I-progesterone with and without the addition of 200 fold excess of unlabeled progesterone. The final volume of the mixture was completed with TEMG buffer pH 7.4 to 1 ml, the mixtures were incubated for 16 hrs. at 4°C. At the end of a period of incubation, bound and unbound progesterone were separated by charcoal adsorption with a suspension of dextran-coated. For this purpose 250 µl of dextran-coated charcol (DCC) were added. The tubes were shaken for 10 min and then centrifuge for 10 min at 2000 xg at 4°C. One milliliter of each supernatant was taken and counted by γ-counter. It represents nuclear bound progesterone.

Solutions

All solutions were prepared as described previously in section (2.1.7).

Calculations

1. The counted radioactivity in each tube (expressed in cpm) represents the total binding (TB).

2. The counted radioactivity (expressed in cpm) in the tubes contained ^{125}I-progesterone and excess of unlabeled progesterone represent the non-specific binding (NSB).

3. The specific binding (SB) was calculated by subtracting the radioactivity (cpm) obtained in the presence of unlabeled progesterone from that produced in the absence of unlabeled progesterone : -

$$SB \text{ (cpm)} = TB \text{ (cpm)} - NSB \text{ (cpm)}$$

4. The percent of specific binding (SB%) can be calculated from the following formula : -

$$SB\% = \frac{SB\ (cpm)}{TC\ (cpm)} \times 100$$

Where :-

SB = Specific binding (cpm)

TC = Total count of ^{125}I-progesterone (cpm) used in each tube.

2.5 Radioreceptor studies of ^{125}I-progesterone binding to its receptors in human benign and malignant ovarian tumors

All the following experiments were carried out with two sets of duplicate tubes. The first one was used to estimate the total binding and the second to estimate the non-specific binding.

2.5.1 The effect of different concentrations of progesterone receptors on the binding of ^{125}I-progesterone in ovarian tumor homogenate

One hundred microliters (6.23 nM) of ^{125}I-progesterone were added to 200 μl of increasing amount (50,100,150,200,250 μg) of (benign and malignant) cytosol fraction in a final volume of 1 ml (completed with TEMG buffer pH 7.4) with and without the addition of 200 fold excess of unlabeled progesterone. At the end of incubation (16 hrs) at 4°C, the bound progesterone estimated by adding 250 μl of DCC, then the tubes were shaken for 10 min and centrifuged for 10 min at 4°C at 2000 xg. One milliliter of each supernatant was taken and counted by γ-counter. It represents the bound hormone.

Solutions

All solutions were prepared as described previously in section (2.1.7).

Calculations

1. The percent of specific binding (SB%) was calculated according to the formula mentioned in section (2.4.2).

2. The percent of specific binding was plotted against the amount of protein receptors incubated in each mixture.

2.5.2 The effect of different concentrations of ^{125}I-progesterone on the binding with its receptors in benign and malignant ovarian tumor homogenate

Increasing concentration (2.75-15.58 nM) of ^{125}I-progesterone was each added to 200 μl (250 μg protein) of (benign and malignant) cytosol fraction in first set of tubes with a final volume of 1 ml (complete with TEMG buffer pH 7.4). The second set of tubes consists of the same reactant plus 200 fold excess of unlabeled progesterone. After incubation for 16 hrs at 4°C, 250 μl of DCC were added in order to estimate the bound progesterone, then the tubes were shaken for 10 min and centrifuge for 10 min at 4°C at 2000 xg. One milliliter was taken from each supernatant was taken and counted by γ-counter. It represents the bound progesterone.

Solutions

All solutions were prepared as described previously in section (2.1.7).

Calculations

1. The same mathematical formula mentioned in section (2.4.2) was used to calculate (SB%).

2. The value of (SB%) was plotted against the concentration of ^{125}I-progesterone.

2.5.3 The effect of different pH on the binding of ^{125}I-progesterone to the receptors in ovarian tumor homogenate.

Two hundred microliters (250 μg protein) of (benign and malignant) cytosol fraction were added to 150 μl (9.345 nM) of ^{125}I-progesterone with or without the addition of 200 fold excess of unlabeled progesterone. The volume of the mixtures were made up to 1 ml with TEMG buffer of different pHs (6.8, 7.0, 7.2, 7.4, 7.6, 7.8, 8.0 and 8.2). The tubes were incubated at 4°C for 16 hrs. After incubation, the bound progesterone was estimated as mentioned in section (2.4.2).

Solutions

All solutions were prepared as described previously in section (2.1.7).

Calculations

1. The (SB%) was estimated as mentioned in section (2.4.2) at each pH.

2. The percent of specific binding (SB%) was plotted against their corresponding pH.

2.5.4 Temperature dependency of the binding between ^{125}I-progesterone and their receptors in ovarian tumor homogenate.

One hundred and fifty microliters (9.345 nM) of ^{125}I-progesterone was added to 200 μl (250 μg protein) of (benign and malignant) cytosol fraction in a final volume of 1 ml (completed with TEMG buffer pH 7.4 for benign tumor and pH 7.6 for malignant tumor) with and without the addition of 200 fold excess of unlabeled progesterone. After incubation 16 hrs at 4°C, the bound progesterone was estimated as mentioned in section (2.4.2).

The experiment was performed at different temperatures (25, 37 and 45°C).

Solutions

All solutions were prepared as described previously in section (2.1.7).

Calculations

1. The percent of specific binding (SB%) was estimated according to section (2.4.2) at each temperature.

2. The percent of specific binding (SB%) was plotted against the different temperatures of incubation.

2.5.5 The choice of most appropriate incubation time of ^{125}I-progesterone binding to receptors in ovarian tumor homogenate.

One hundred and fifty microliters (9.345 nM) of ^{125}I-progesterone were added to 200 μl (250 μg protein) of benign and malignant cytosol fraction in a final volume of 1 ml (completed with TEMG buffer pH 7.4 for benign tumor and pH 7.6 for malignant tumor) with and without the addition of 200 fold excess of unlabeled progesterone. The tubes were incubated at (25°C for benign tumor and 4°C for malignant tumor) at different time intervals (2, 4, 6, 8, 12, 16, 18 and 20 hrs). At the end of incubation, the bound hormone was estimated as mentioned in section (2.4.2).

Solutions

All solutions were prepared as described previously in section (2.1.7).

Calculations

1. The (SB%) was estimated as mentioned in section (2.4.2) at each incubation time.

2. The percent of specific binding (SB%) was plotted against their corresponding time.

2.5.6 Competitive effect of different concentrations of unlabeled progesterone, estradiol and testosterone on the binding of ^{125}I-progesterone to its receptors in human ovarian tumors

The experiment was carried out at the optimum conditions of ^{125}I-progesterone concentration (9.345 nM), protein receptor concentration (250 μg), temperature, pH and time of incubation (25°C, pH 7.4 and 6 hrs for benign tumor) and (4°C, pH 7.6 and 18 hrs for malignant tumor). The experiment was performed by adding 150 μl of ^{125}I-progesterone to 200 μl (250 μg protein) of crude cytosol fraction with and without the addition of increasing concentrations (50-400 ng/ml) of unlabeled progesterone in a final volume of 1 ml (completed with TEMG buffer). After incubation, the bound hormone was estimated as mentioned in section (2.4.2). The experiment was repeated with increasing concentration of unlabeled estradiol and testosterone.

Solutions

All solutions were prepared as described previously in section (2.1.7).

Calculations

1. The percent of relative specific binding (RSB%) was estimated from the following formula:-

$$RSB\% = \frac{\text{Specific binding of }^{125}\text{I-progesterone in the presence of competitor}}{\text{Specific binding of }^{125}\text{I-progesterone in the absence of competitor}} \times 100$$

2. The percents of relative specific binding (RSB%) were plotted against the different concentrations of competitors (progesterone, estradiol and testosterone).

2.5.7 The effect of different halides on the binding of ^{125}I-progesterone to its receptors in ovarian tumor homogenate

Two hundred microliters (250 μg protein) of (benign and malignant) crude cytosol fraction were incubated with 150 μl (9.345 nM) of ^{125}I-progesterone with and without the addition of 200 fold excess of unlabeled progesterone. Add 50 μl of (2M) NaF, the final volume of 1 ml (completed with TEMG buffer pH 7.4 for benign tumor and pH 7.6 for malignant tumor). The final concentration of the adding halide (0.1M). The tubes were incubated (6 hrs at 25°C for benign tumor and 18 hrs at 4°C for malignant tumor), then bound progesterone was estimated as mentioned in section (2.4.2). The control was used without the addition of any halide.

The experiment was performed by using different halides at final concentration 0.1M (NaCl, NaBr and NaI).

Solutions

Halides solutions were prepared in concentration of (2M) in TEMG buffer pH 7.4 (benign tumor) and pH 7.6 (malignant tumor):-

• NaF : 0.8398 gm of NaF in 10 ml of TEMG buffer.

• NaCl : 1.1688 gm of NaCl in 10 ml of TEMG buffer.

• NaBr : 2.0578 gm of NaBr in 10 ml of TEMG buffer.

• NaI : 2.9978 gm of NaI in 10 ml of TEMG buffer.

Calculations

1.The (SB%) was estimated according to section (2.4.2) at each halide.

2.The percent of specific binding (SB%) was plotted against each type of halide.

2.5.8 The effect of monovalent salts on the binding of ^{125}I-progesterone to its receptors in benign and malignant human ovarian tumors

The experiment was performed at optimum conditions (temperature, time, pH, ^{125}I-progesterone and protein concentration) as mentioned in section (2.5.7) with one exception that the addition of 50 μl (2M) from different monovalent salts (NH_4Cl, KCl, LiCl, CsCl and NaCl), the final concentration of the adding salts (0.1M). The bound hormon was estimated as mentioned in section (2.4.2). The control was used without the addition of any salts.

Solutions

Monovalent salts were prepared in (2M) in TEMG buffer (pH 7.4 for benign tumor and pH 7.6 for malignant tumor) :

• NH_4Cl : 1.0689 gm in 10 ml of TEMG buffer.

• KCl : 1.491 gm in 10 ml of TEMG buffer.

• LiCl : 0.8478 gm in 10 ml of TEMG buffer.

• CsCl : 3.3672 gm in 10 ml of TEMG buffer.

• NaCl : 1.1688 gm in 10 ml of TEMG buffer.

Calculations

1. The (SB%) was estimated according to section (2.4.2) at each salt type.

2. The percent of specific binding (SB%) was plotted against each salt type.

2.5.9 The effect of divalent cations on the binding of ^{125}I-progesterone with its receptors in ovarian tumor homogenate

To evaluate the effect of divalent cations on the progesterone binding with its receptors in human ovarian homogenate, the experiment was carried out at optimum conditions of time, temperature, pH, ^{125}I-progesterone and protein concentration as mentioned in section (2.5.7) with one exception that the addition of 50 μl (500 mM) from different divalent cations ($MgCl_2.6H_2O$, $MnCl_2.4H_2O$, $CuSO_4.5H_2O$, $CaCl_2.2H_2O$ and $ZnCl_2$), the final concentration of the adding salts (25 mM). The bound progesterone was estimated as mentioned in section (2.4.2). The control was used without the addition of any salt.

Solutions

The stock solutions (500 mM) of divalent salts were prepared as the following :

- $MgCl_2.6H_2O$: 1.01655 gm in 10 ml of TEMG buffer.
- $MnCl_2.4H_2O$: 0.98955 gm in 10 ml of TEMG buffer.
- $CuSO_4.5H_2O$: 1.2484 gm in 10 ml of TEMG buffer.
- $CaCl_2.2H_2O$: 0.7351 gm in 10 ml of TEMG buffer.
- $ZnCl_2$: 0.6814 gm in 10 ml of TEMG buffer.

Calculations

1. The (SB%) was estimated according to section (2.4.2) for each salt.

2. The percent of specific binding (SB%) was plotted against salt type.

2.5.10 The effect of Urea on the binding of ^{125}I-progesterone with its receptors in benign and malignant ovarian tumor

The experiment was carried out at optimum conditions of time, temperature, pH, ^{125}I-progesterone and protein concentration with one exception that the reaction mixtures were completed to 1 ml with TEMG buffer containing urea ranging in their concentration from (0.25 to 4 M). The bound progesterone was estimated as mentioned in section (2.4.2). The control in this experiment was used without the addition of urea.

Solutions

1. The stock urea solution (8M) was prepared by dissolving (48 gm) in (100 ml) of TEMG buffer pH 7.4 (benign tumor) and pH 7.6 (malignant tumor).

2. Various solutions of urea ranging in their molar concentrations from 0.25 to 4M were prepared by the serial dilution of the stock.

Calculations

1.The (SB%) was estimated using the same mathematical formula of section (2.4.2) for each urea concentration.

2.The values of (SB%) was plotted versus the corresponding molar concentrations of urea solutions.

2.5.11 The effect of polyethylene glycol on the binding of ^{125}I-progesterone with their human ovarian tumors receptors

The experiment was carried out at optimum conditions of time, temperature, pH, ^{125}I-progesterone and protein concentration with an exception that the reaction mixtures were completed to 1 ml with TEMG buffer containing various percent of polyethylene glycol (PEG-10000) ranging from (2 to 10%). The bound progesterone was estimated as mentioned in section (2.4.2). The control used in this experiment was without the addition of PEG.

Solutions

The stock solution of PEG-10000 was prepared by dissolving 20 gm of PEG in 100 ml of TEMG buffer (pH 7.4 for benign tumor and pH 7.6 for malignant tumor).

2, 4, 6, 8 and 10% of PEG solutions were prepared by an appropriate serial dilution of the stock solution.

Calculations

1. The (SB%) was estimated as mentioned in section (2.4.2) at each PEG percent.
2. The (SB%) values were plotted against PEG precent.

2.6 Kinetics and thermodynamics of the interaction of progesterone with its cytosolic receptors

2.6.1 The time course of ^{125}I-progesterone binding to its receptors in benign and malignant ovarian tumors

1. At zero time, the experiment was carried out at the optimum conditions of pH, ^{125}I-progesterone and protein concentration. Incubation was carried out for several time intervals (2, 4, 6, 8, 12, 16, 18 and 20 hrs).

2. After each time intervals, the bound progesterone was estimated as described previously in section (2.4.2).

3. Parallel experiments were performed to determine the amount of non-specific binding.

4. To determine the time course of the association of ^{125}I-progesterone with its receptors in benign and malignant ovarian tumors at different temperatures, the above experiment was performed at four temperatures (4, 25, 37 and 45°C).

Solutions

All solutions were prepared as described previously in sections (2.1.7).

Calculations

1. The value of ^{125}I-progesterone bound specifically in (nanomole of ^{125}I-progesterone per mg of protein) was calculated according to the following formula :

$$
\begin{bmatrix} \text{The value of specifically} \\ \text{bound }^{125}\text{I-progesterone} \\ \text{(nmole/mg protein)} \end{bmatrix} = \frac{\begin{bmatrix} \text{Specifically bound} \\ ^{125}\text{I-progesterone (nM)} \end{bmatrix} \times \begin{bmatrix} \text{Incubation volume} \\ \text{in (liter)} \end{bmatrix}}{\text{mgs of protein in incubation medium}}
$$

$$
\begin{bmatrix} \text{Specifically bound} \\ ^{125}\text{I-progesterone (nM)} \end{bmatrix} = \frac{\begin{bmatrix} \text{Total binding} \\ \text{(cpm)} \end{bmatrix} - \begin{bmatrix} \text{Non-specific} \\ \text{binding (cpm)} \end{bmatrix}}{\text{Total count (cpm)}} \times \begin{bmatrix} \text{Total} \\ \text{concentration of} \\ ^{125}\text{I-progesterone} \\ \text{in incubation} \\ \text{medium} \end{bmatrix}
$$

$$
\begin{bmatrix} \text{The precent of} \\ \text{specific binding} \\ \text{(SB\%)} \end{bmatrix} = \begin{bmatrix} \dfrac{\left[\text{Total binding (cpm)}\right] - \left[\text{Non-specific binding (cpm)}\right]}{\text{Total count (cpm) of }^{125}\text{I-progesterone used in each tube}} \end{bmatrix} \times 100
$$

2. The percent of specific binding was plotting against the time of incubation.

2.6.2 Determination of the concentration of progesterone receptors and the affinity constant of ^{125}I-progesterone association with its receptors in benign and malignant ovarian tumors

Crude cytosol progesterone receptors were measured by the addition of increasing concentration (3.115-9.345 nM) of ^{125}I-progesterone to 200 μl (250 μg protein) with and without the addition of 200 fold excess of unlabeled progesterone in a final volume of 1 ml (complete with TEMG buffer pH 7.4 for benign tumors and pH 7.6 for malignant tumors). The tubes incubated for 6 hrs at 25°C (benign tumors) and 18 hrs at 4°C (malignant tumors) in order to attain an equilibrium state. The bound progesterone was estimated as mentioned in section (2.4.2). All previous steps of this experiment were performed at different temperatures. The time of incubation needed to get the equilibrium state at each

temperature was obtained from the related time course pattern.

Solutions

All solutions were prepared as described previously in section (2.1.7).

Calculations

1. The values of ^{125}I-progesterone which is bound specifically in nonomolar
were calculated using the following formula :

$$B = \frac{\text{Total binding - Nonspecific binding}}{\text{Total count}} \times \left[\begin{array}{c} \text{Concentration } ^{125}\text{I-progesterone (nM)} \\ \text{in each assay tube} \end{array} \right]$$

2. The concentration of receptors and the affinity constant were determinatied
according to Scatchard equation [178,179].

$$\frac{B}{F} = \frac{1}{Kd} (B_{max} - B)$$

$$Ka = \frac{1}{Kd}$$

Where :

B : The concentration of specifically bound progesterone.

F : The concentration of free progesterone.

Ka : The affinity constant.

B_{max} : The maximal binding capacity.

Kd : The dissociation constant.

3. The B/F values were plotted against the values of the B, the receptor
concentration and the affinity constant were calculated from the x-axis and
the slope of the straight line respectivily.

2.6.3 The thermodynamics of ^{125}I-progesterone interaction with its receptors in benign and malignant ovarian tumors

Two hundred microliters of ovarian tumor homogenate (250 μg protein) were incubated with (9.345 nM) of ^{125}I-progesterone at 25°C for 6 hrs (for benign tumors) and at 4°C for 18 hrs (for malignant tumors). The final volume 1 ml (completed with TEMG buffer pH 7.4 for benign tumors and pH 7.6 for malignant tumors). The steps 2,3 and 4 of the experiment (2.6.1) were carried out for several time intervals (2, 4, 6, 8, 12, 16, 18 and 20 hrs).

Solutions

All solutions were prepared as described previously in section (2.1.7).

Calculations

1. The thermodynamic parameters of standard state were obtained from Van't Hoff plot, the values of the natural logarithm of equilibrium constant (affinity constant Ka) obtained at different temperatures were plotted against the reciprocal values of absolute temperature in Kelvin (1/T), according to the following equation :

$$\ln Ka = \frac{\Delta S^{o}}{R} - \frac{\Delta H^{o}}{RT}$$

Where :

ΔH^{o} : The enthalpy change of the standard state.

ΔS^{o} : The entropy change of the standard state.

ΔR : The gas constant (8.3144 J.mole^{-1}.K^{-1}).

H^{o} value was obtained from the slope of the linear relationship of the plot.

The change in Gibbs free energy of the standard state (ΔG^{o}) was calculated from the following equation :

$$\triangle G^{o} = -RT \ln Ka$$

In addition, the standard state entropy change (ΔS^o) was calculated from the following formula :

$$\Delta S^o = \frac{(\Delta H^o - \Delta G^o)}{T}$$

2. The thermodynamic parameters of the transition state were estimated from Arrhenius plot of $\ln K_{+1}$ versus $(1/T)$ which gives a linear relationship according to the following equation :

$$\ln K_{+1} = \ln A - \left[\frac{Ea}{RT} \right]$$

Where :

     A : The Arrhenius constant,

     Ea : The activation energy,

     R : The gas constant, and

     T : Absolute temperature.

The Activation energy of the binding reaction was calculated from the slop of the straight line. The enthalpy of the transition state (ΔH^*) was determined from :

$$\triangle H^* = Ea - RT$$

The transition state of free energy (ΔG^*) was calculated from the following equation :

$$\triangle G^* = - RT \ln K_{+1} + RT \ln \left(\frac{KT}{h} \right)$$

Where :

     K : is Boltzmann constant (1.38×10^{-23} J.deg^{-1}).

     h : is Plank constant (0.662×10^{-33} J.sec^{-1}).

The change in entropy of the transition state ($\triangle S^*$) was calculated from the following formula :

$$\triangle S^* = \frac{(\triangle H^* - \triangle G^*)}{T}$$

2.7 Isolation of cytosol progesterone receptors using gel filtration technique

2.7.1 Gel filtration and column packing [180,181]

The gel (sephadex G-150) was allowed to swell in excess of buffer A pH 7.4 (50 ml buffer/g of gel) and left to stand for three days (72 hrs) at room temperature without stirring, the gel slurry were degassed by section for 30 min, then the swollen gel was poured carefully into a vertical glass-column down the wall using a glass-rod. After the gel has settled the column was equilibrated with buffer (A) pH 7.4 for 24 hrs with the dimension of (0.7x23 cm).

2.7.2 Void volume (Vo) determination

The void volume of the column was estimated using blue dextran 2000 with concentration of 2 mg/ml dissolved in buffer (A) pH 7.4, 1 ml of blue dextran solution was applied to the column carefully, then elution was carried out with the same buffer using a flow rate of 6 ml/hr. Fractions of 1 ml were collected and their absorbances were measured at 600 nm. The volume of the buffer that required to elute the blue dextran represents the void volume (5 ml) of the column.

2.7.3 Isolation procedure

One milliliter of crude cytosol (3.5 mg protein) was applied to the surface of sephadex G-150 column (0.7x23 cm) equilibrated with buffer (A). The sample was eluted using the same buffer, fractions of 1 ml were collected at a flow rate of 6 ml/hr. The absorbances of the fractions collected were measured at 280 nm and the protein contents were determined by the method of lowry *et al.*[176].

2.7.4 The preliminary test of the binding of ^{125}I-progesterone to the isolated fractions separated by gel filtration

Two hundred microliters of isolated fractions were added to 150 μl (9.345 nM) of ^{125}I-progesterone with and without the addition of 200 fold excess of unlabeled progesterone in a final volume of 1 ml completed with TEMG buffer. The tubes were incubated for (6 hrs at 25°C for benign tumors) and (18 hrs at 4°C for malignant tumors). The bound progesterone was measured as described in section (2.4.2).

Solutions

Buffer (A) : TEMG buffer pH 7.4 (was prepared as described previously in section (2.1.7) containing 0.02% sodium azide.

Calculations

1. The dimensions of the column were chosen according to the following equations [180] :

$$\text{Diameter (cm)} = \sqrt[3]{\frac{m}{10}}$$

Where :

  m : is the amount of protein in mg.

    Length (cm) = 30 x diameter

2. The values of SB% for the eluted fractions were calculated in the same method as that of the previous experiments.

3. The value of SB% and absorbances at 280 nm were plotted against the fraction number.

4. The isolation fold for each progesterone receptor for benign and malignant human ovarian tumors was estimated from the following formula :

$$\text{Isolation fold} = \frac{\text{Specific binding of isolated receptor (nmole/mg protein)}}{\text{Specific binding of crude receptor (nmole/mg protein)}}$$

2.7.5 Determination of the molecular weight by gel filtration chromatography

The same sephadex G-150 column used in section (2.7.1) was calibrated for molecular weight determination.

Standard proteins listed in table (2-5) were applied separately into the sephadex column (2 mg/ml). Elution was carried out with buffer (A) at a flow rate of 6 ml/hr. The absorbances of the fractions collected (1 ml each fraction) were measured at 280 nm to evaluate the elution volume. Receptor preparation was applied to column and eluted at the same conditions. The partition coefficients (K_{av}) of the standard proteins were determined using the folloeing formula :

$$K_{av} = \frac{V_e - V_o}{V_t - V_o}$$

Where :

V_e = Elution volume

V_o = Void volume

V_t = The volume of the bed gel

Table (2-5) : Standard proteins for estimation *M.wt* by gel filtration.

Protein	M.wt "Dalton"
Ferritin	440.000
Catalase	232.000
Aldolase	158.000
BSA	67.000

The values of K_{av} were plotted vs. the values of log *M.wt* of standard proteins. The *M.wt* of progesterone receptor was calculated from the standard curve obtained.

Solutions

Buffer (A) : TEMG buffer pH 7.4 (was prepared as described previously in section (2.1.7) containing 0.02% sodium azide.

2.8 Spectroscopic studies of different isolated forms of progesterone receptors

2.8.1 The U.V. Spectra of isolated progesterone receptors from human benign and malignant ovarian tumors

Two hundred microliters (350 μg protein) of each isolated receptor was completed to 1 ml with distilled water pH 7.4, then placed in a cuvette in sample beam and the absorption spectrum was immediately measured against the adjusted pH distilled water as a reference.

2.8.2 Factors affecting the absorption properties of isolated progesterone receptors in human benign and malignant ovarian tumors

- *pH effect*

Two hundred microliters (350 μg protein) of isolated receptor were completed to 1 ml with distilled water of different pH (2.7, 7.4 and 10.7) then each of which was placed in the test cell and the adjusted pH distilled water was placed in the reference cell and the absorption spectra of different isolated receptors were measured immediately.

- *Polarity effect*

a. The effect of 20% ethanol on the progesterone receptors spectra :

Two hundred microliters (350 μg protein) of isolated receptor were completed to 1 ml with distilled water contains 20% ethanol at pH 7.4 then each of which was placed in the test cell and the 20% ethanol adjusted pH was placed in the reference cell. The absorption spectrum of each sample was measured immediately.

b. The effect of 20% ethylene glycol on the progesterone receptors spectra :

Two hundred microliters (350 μg protein) of isolated receptor were completed to 1 ml with distilled water contains 20% ethylene glycol at pH 7.4 then each of which was placed in the test cell and the 20% ethylene glycol adjusted pH was placed in the reference cell. The absorption spectrum of each sample was measured immediately.

c. The effect of 20% DMSO on the progesterone receptors spectra :

Two hundred microliters (350 μg protein) of isolated receptor were completed to 1 ml with distilled water contains 20% DMSO at pH 7.4 then each of which was placed in the test cell and the 20% DMSO adjusted pH was placed in the reference cell. The absorption spectrum of each sample was measured immediately.

d. The effect of 20% Urea on the progesterone receptors spectra :

Two hundred microliters (350 μg protein) of isolated receptors were completed to 1 ml with distilled water at pH 7.4 containing 20% urea then placed in the test cell against the 20% urea adjusted pH in the reference cell. The absorption spectra of different isolated receptors were measured immediately.

2.8.3 Spectrophotometric pH titration of isolated progesterone receptors in human benign and malignant ovarian tumors

A series of isolated receptors (350 μg protein in 200 μl) were completed to 1 ml with distilled water at pH ranging from (9.0 to 12.5). The maximum absorbance of each sample was measured at a wavelength of 295 nm, the absorbance of λ_{max} at each pH value was plotted versus the corresponding pH.

A nother series of isolated receptors were completed to 1 ml with distilled water at pH ranging from (4.0 to 8.0). The maximum absorbance of each sample was measured at a wavelength of 211 nm, the absorbance of λ_{max} at each pH value was plotted versus the corresponding pH.

2.8.4 The U.V. spectra of ^{125}I-progesterone and the different ^{125}I-progesterone – receptor complexs

2.8.4.1 The U.V. spectra of different human progesterone – receptors complexes of benign and malignant ovarian tumors

The binding experiment of different isolated progesterone receptors with ^{125}I-progesterone was carried out at optimum conditions as explained previously in section (2.5.6). One milliliter of the ^{125}I-progesterone — receptor complex supernatant of each type of isolated receptors was placed in cuvette in the sample beam and the absorption spectrum was measured immediately against an appropriate blank in the reference beam.

2.8.4.2 *The UV. spectrum of ^{125}I-progesterone*

One milliliter of ^{125}I-progesterone was placed in a cuvette in the sample beam and the absorption spectrum was measured immediately against an appropriate blank in the reference beam.

2.9 Statistical analysis

The result of serum progesterone was analyzed statistically and values were expressed as mean ± SD. The levels of significance were determined by student's t-test [182].

CHAPTER THREE

RESULTS & DISCUSSION

RESULTS AND DISCUSSION

3.1 Determination of progesterone levels in sera of ovarian tumors patients

Progesterone levels in sera of patients with benign ovarian tumors (group I) and (pre- and post- menopausal) malignant ovarian tumors (group II, III) were measured by radioimmunoassay (RIA) [183,185]. Three groups were matched with a group of control subject. Table (3-1) shows the results obtained from this study.

The level of serum progesterone in benign ovarian tumor patients was found to be (1.43 $\mp$ 0.41ng/ml) whereas that of (pre- and post- menopausal) malignant ovarian tumor patients were found to be (1.38$\mp$0.54 ng/ml), (0.074$\mp$0.022 ng/ml) respectively.

Table (3-1): serum progesterone levels (ng/ml) in patients with benign and malignant ovarian tumors. Details are described in section (2.2).

Group	Patients	No. of cases	Age (year)	Serum progesterone (ng/ml)
I	Pre-menopausal of benign ovarian tumors	14	30-37	1.43 ± 0.41
II	Pre-menopausal of malignant ovarian tumor	8	29-35	1.38 ± 0.54
	Control	16	28-38	0.95 ± 0.21
III	Post-menopausal of malignant ovarian tumor	6	57-62	0.074 ± 0.022
	Control	10	55-65	0.039 ± 0.018

Student's T-test analysis has revealed that there is highly significant increase of serum progesterone levels in benign tumors ($p<0.0005$) and (pre- and post- menopausal) malignant tumor ($p<0.005$). These results are nearly similar to those obtained previously by other investigator [186-188].

3.2 Binding studies of ^{125}I-progesterone with it's receptors in human benign and malignant ovarian tumors

3.2.1 Preliminary test of ^{125}I-progesterone binding with its receptors in ovarian tumors homogenates

Benign and malignant ovarian tumors patients (pre-menopausal women) were investigated to evaluate the presence of progesterone receptors after removal of tumors surgically. Cytosolic progesterone receptor were detected through the incubation of ^{125}I-progesterone with crude cytosol and the bound hormone was separated by dextran-coated charcol method and then measured by γ-counter. The preliminary conditions used in this experiment resulted in (3.2%) and (4%) specific binding for benign and malignant tumors homogenate, respectively.

Several studies on the presence of progesterone receptors in ovarian tumors have been done [189-191].

The data obtained in this preliminary experiment have revealed the presence of cytosolic progesterone receptors in human benign and malignant ovarian tumors.

3.2.2 Radio Receptor Assay (RRA) studies of ^{125}I-progesterone binding with it's receptors in benign and malignant ovarian tumors homogenates

3.2.2.1 The effect of different concentrations of progesterone receptors on the binding of ^{125}I-progesterone in ovarian tumor homogenates

To determine whether the specific binding was proportional to the amount of progesterone receptors used, increasing amount of cytosol homogenate were incubated with ^{125}I-progesterone with and without the addition of unlabelled progesterone, according to the details in section (2.5.1). The relationship observed from figure (3-1) shows the increasing values of the specific binding percent (SB%) with increasing amount of receptor protein, until a point of maximum binding, then a resultant decreases in the specific binding percent.

These results indicate that progesterone receptors binding principally depende on the amount of receptor protein in the reaction mixture [192].

In all the subsequent experiments, 250 μg of receptor protein of benign and malignant tumors homogenate were used.

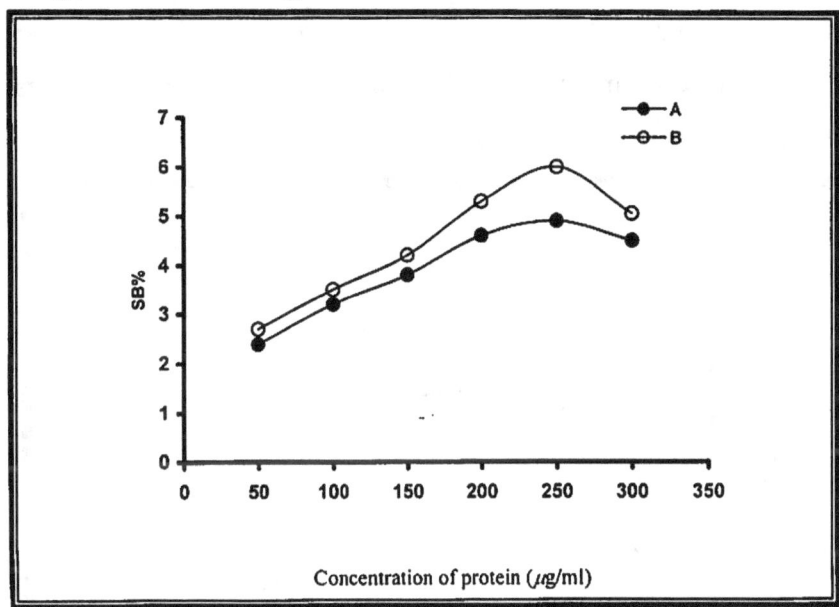

Figure (3-1): The effect of different protein concentrations in:
- A: Benign ovarian tumor homogenate.
- B: Malignant ovarian tumor homogenate.
- with [125] I-progesterone.
- Details are described in section (2.5.1)

3.2.2.2 The effect of different concentrations of ^{125}I-progesterone on the binding with its receptors in benign and malignant ovarian tumors homogenates

One of the most important criteria of the hormone receptors is its saturability. To fulfil this criterion and to estimate the suitable concentration of ^{125}I-progesterone, the experiment was carried out in the presence of 250 μg of benign and malignant receptors and increasing concentration of ^{125}I-progesterone as mentioned previously in section (2.5.2). Figure (3-2) is a representative of ^{125}I-progesterone binding curve with its receptors in benign and malignant tumors homogenate. It is shown that the specific binding of ^{125}I-progesterone with receptor protein binding is a saturable process but complete saturation however is theoretically never reached unless the amount of steroid hormone used reached infinity [(193)].

As shown in figure (3-2), the receptor was saturated with ^{125}I-progesterone at the concentration of (9.345 nM). Accordingly, in all the subsequent experiments, (9.345 nM) per (150 μl) of ^{125}I-progesterone was used in the incubation mixture for both benign and malignant tumors homogenate.

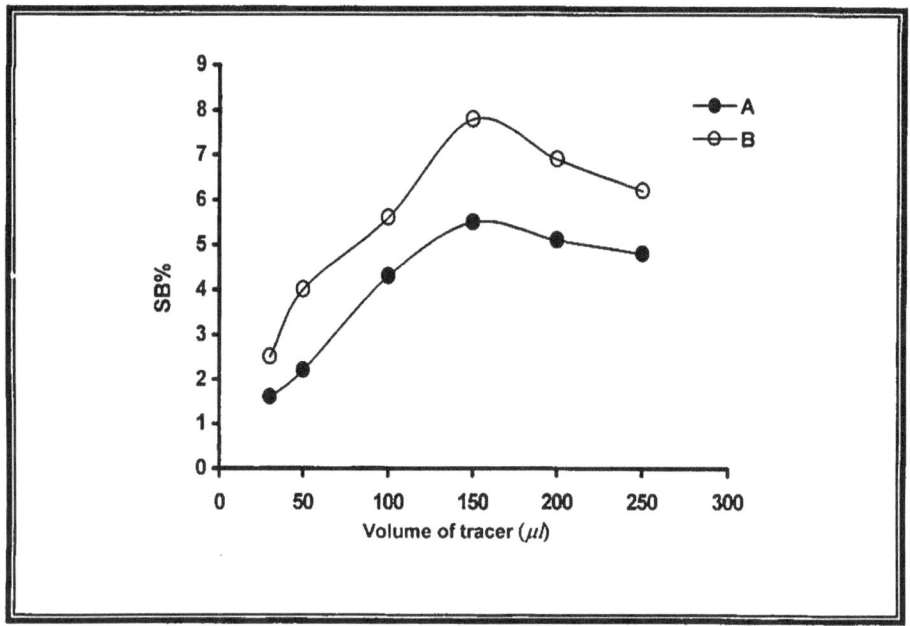

Figure (3-2): Effect of different concentrations of [125] I-progesterone on the binding with:
A: Benign ovarian tumor homogenate.
B: Malignant ovarian tumor homogenate.
Details are described in section (2.5.2)

3.2.2.3 The effect of different pH on the binding of ^{125}I-progesterone to the receptors in ovarian tumors homogenates

The effect of pH on the specific binding of ^{125}I-progesterone to its receptors in benign and malignant tumors homogenate was investigated. Figure (3-3) shows that the optimum pH was found to be (7.4) for the binding of benign receptors and (7.6) for the binding of malignant receptors. The same figure shows a decreasing in specific binding percent at the pH higher or lower than the optimum pH. These results indicate that the binding was pH-dependant and the shift in the pH of the environment may affect the propertied of the macromolecules involved in the binding. This effect includes the induction of protonation-deprotonation processes occurring with the ionizable groups of the amino acids present in the binding domain of the macromolecules [194,195].

A previous work on the rat ovary, done by RadioImmunassay, suggested that pH 7.0 is the optimum pH for the binding [196]. While another work suggested that the highest binding was seen at pH (7-7.5) in luteal homogenate from human ovarian corpora lutea [197].

In another study reported that the pH 7.6 is the optimum for the binding of the ^{125}I-progesterone with its receptor in human uterine tumors homogenates [198]. The difference in the pH range may be due to the difference of the homogenates and to the binding conditions used.

According to the results obtained in this analysis, the pH of buffers used in all subsequent experiments were adjusted to 7.4 for benign tumors and 7.6 for malignant tumors.

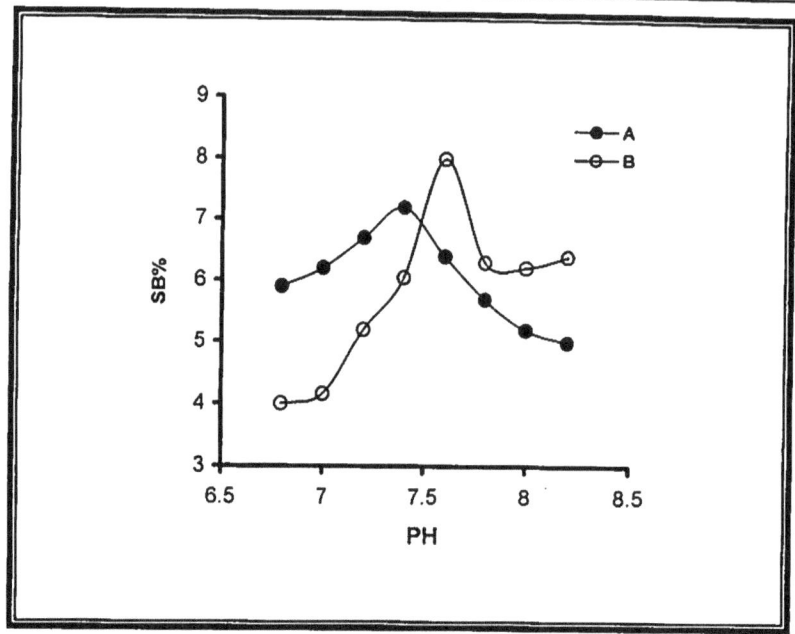

Figure (3-3): pH dependency of ^{125}I-progesterone binding with its receptors in:

    A: Benign ovarian tumor homogenate.
    B: Malignant ovarian tumor homogenate.
    Details are described in section (2.5.3).

3.2.2.4 The effect of temperature on the binding of ^{125}I-progesterone to its receptors in benign and malignant ovarian tumors homogenates

Temperature dependency of the association of ^{125}I-progesterone to its receptors was investigated. Cytosol fraction of benign and malignant ovarian tumors were incubated at different temperature as in section (2.5.4). Figure (3-4) revealed that the specific binding of ^{125}I-progesterone to its cytosol receptors was maximal at 25°C for benign tumors homogenate and its was maximum at 4°C for malignant one. The specific binding was decreased at temperature increase after maximal value of binding. The loss of binding activity may be due to degradation of the receptor [199] and or the irreversible dissociation of the hormone receptor complexes.

Previous studies on the rat ovary, reported that the ɟbinding was temperature dependent and the optimum temperature is 25°C in granulosa cells isolated from rat ovary [200].

According to these results, the temperatures in all subsequent experiments were 25°C for benign tumors and 4°C for malignant tumors.

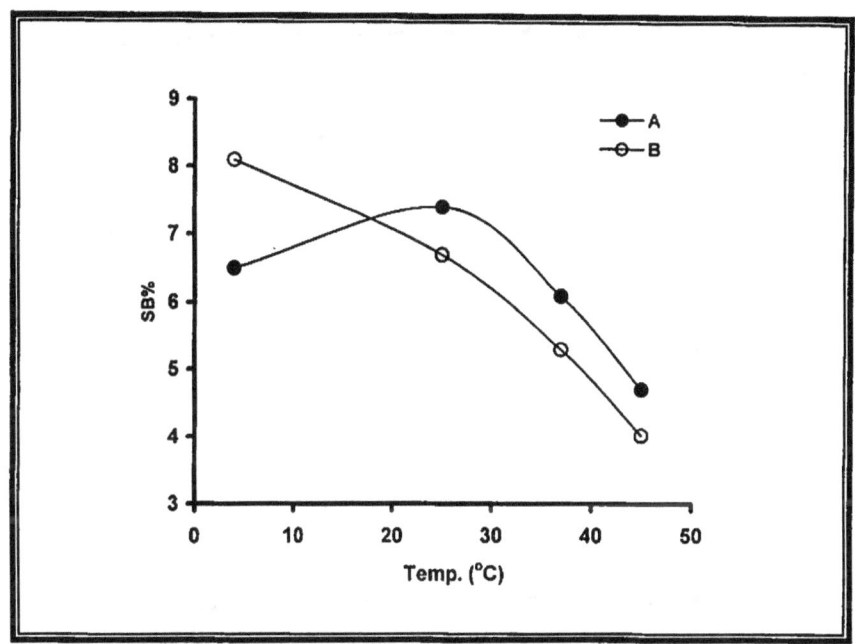

Figure (3-4): Effect of temperature on ^{125}I-progesterone binding with its
            receptors in:
            A: Benign ovarian tumor homogenate
            B: Malignant ovarian tumor homogenate.
            Details are described in section (2.5.4).

3.2.2.5 The choice of most appropriate incubation time of ^{125}I-progesterone binding to its receptors in ovarian tumors homogenate

To choose the most appropriate incubation time at the optimum temperature (25°C for benign tumors and 4°C for malignant tumors) the experiment was carried out at different time intervals (2-20 hrs).

Figure (3-5) shows that the optimal binding of ^{125}I-progesterone with its receptors in benign tumors occurred within (6 hrs) at 25°C and in malignant tumors occurred within (18 hrs) at 4°C.

Previous studies reported that the maximal binding in ovarian homogenate from pseudo pregnant rat was at 25°C for 4 hrs by RIA [201].

According to our results, in all other subsequent experiment was shown that the specific binding of ^{125}I-progesterone to its receptors were maximal at (6 hrs) for benign tumors and (18 hrs) for malignant one.

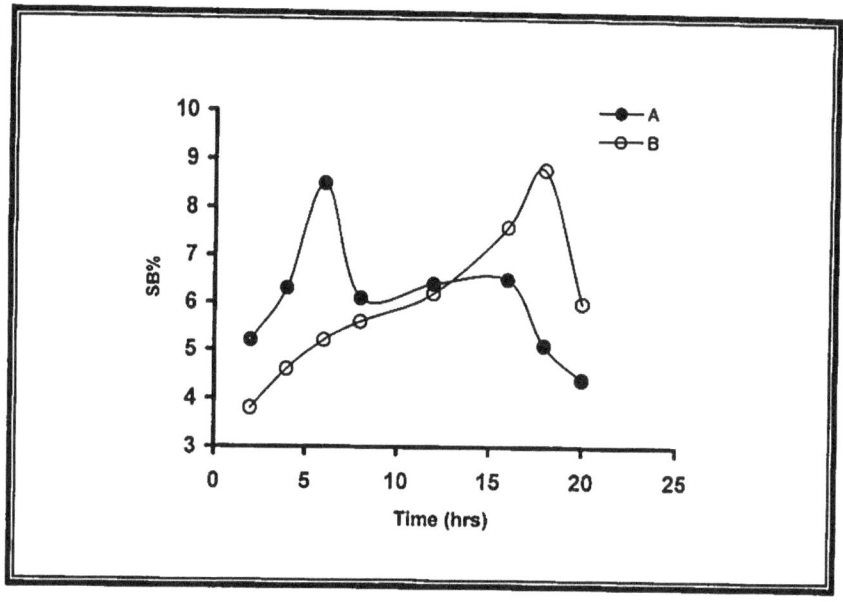

Figure (3-5): Time course of ^{125}I-progesterone binding with:
    A: Benign ovarian tumor homogenate.
    B: Malignant ovarian tumor homogenate.
    Details are described in section (2.5.5).

3.2.2.6 Competitive effect of different concentration of unlabeled progesterone, estradiol and testosterone on the binding of ^{125}I-progesterone to its receptors in benign and malignant ovarian tumors

The specificity of benign and malignant progesterone receptors for binding with different steroids were demonstrated by a decrease in receptor bound radioactivity after incubating the cytosol fraction with increasing amount of unlabeled steroid hormones. Figure (3-6 A&B) shows that the binding of ^{125}I-progesterone to its benign and malignant receptors were effectively inhibited by unlabeled progesterone and with a lesser extant by unlabeled testosterone and estradiol, respectively. The frequency of inhibiting strength of these steroids was found to be as the following :

Progesterone > Testosterone > Estradiol

The results pointed out that the highly potent competitor of ^{125}I-progesterone binding was the unlabeled progesterone.

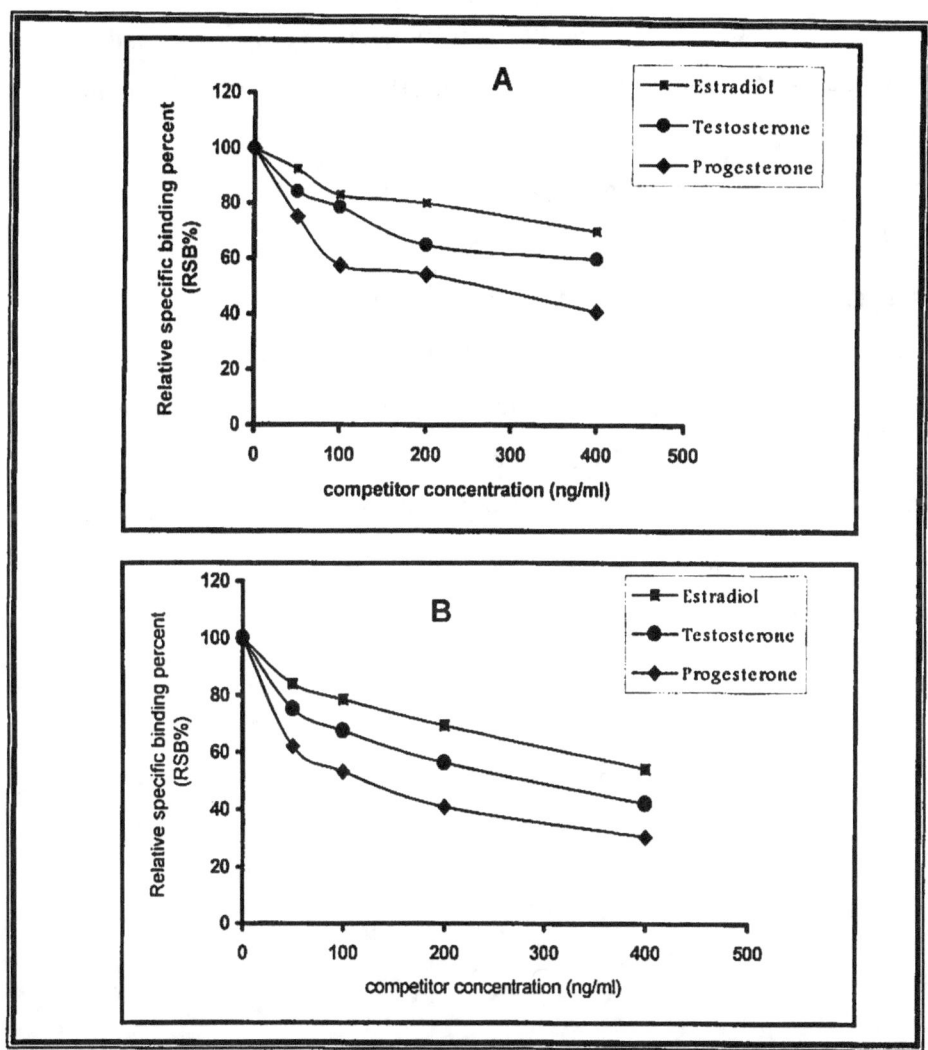

Figure (3-6): The effect of different concentrations of unlabeled Estradiol, Testosterone and progesterone on the binding of [125] I-Progesterone with its receptors in:
A: Benign ovarian tumor homogenate.
B: Malignant ovarian tumor homogenate
Details are described in section (2.5.6)

3.2.2.7 The effect of different halides on the binding of ^{125}I-progesterone to its receptors in benign and malignant ovarian tumors

Different halides of sodium were investigated to study their action on the binding of ^{125}I-progesterone with its receptors in benign and malignant ovarian tumors homogenate. The sodium halides in the incubation mixture caused the increase of specific binding percent with the decrease of halides size, the induced-activation of the percent of specific binding according to the following sequence :

$$NaF > NaCl > NaBr > NaI$$

Melander and Horvath (1977) reported that the effect of halides salt type on hydrophobic interactions is quantified by its molar surface tension increment (MSTI) which is a measure of the increase in surface tension by the salt, also they found that this parameter increase as the following sequence :

$$NaF > NaCl > NaBr > NaI$$

The same researchers found that halides with higher MSTI values will strengthen the hydrophobic interactions while halides with lower MSTI values reverse this effect [202].

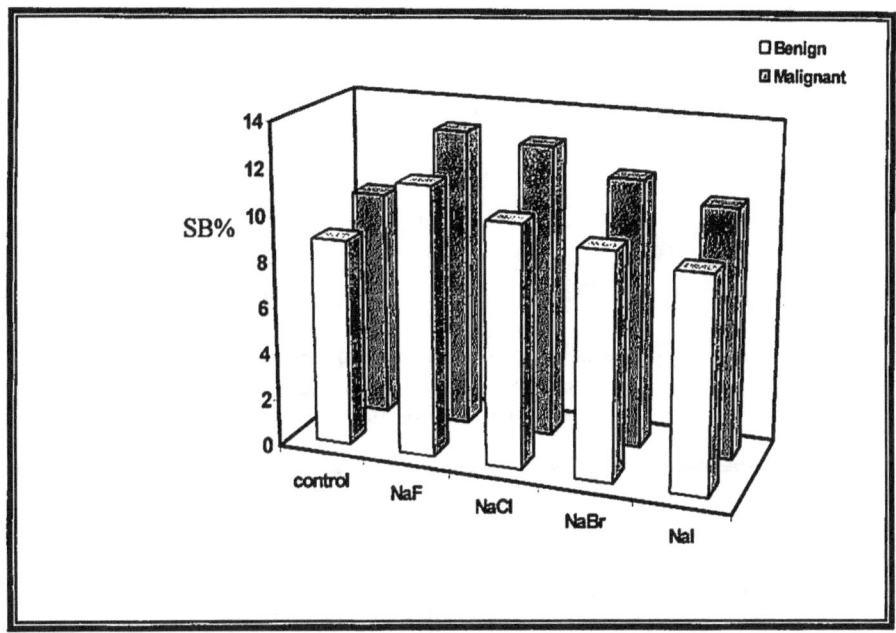

Figure (3-7): The effect of different halides on the binding of ^{125}I-progesterone
with it's receptors in benign and malignant ovarian tumor
homogenate.
Details are described in section (2.5.7).

3.2.2.8 The effect of monovalent and divalent salts on the binding of ^{125}I-progesterone to its receptors in benign and malignant ovarian tumors

Figure (3-8) shows the effect of different salts on the extent of the binding of ^{125}I-progesterone to its receptors in benign and malignant ovarian tumors. $CaCl_2$ at (25 mM) was shown to increase the binding compared with control value in benign and malignant tumors. The other salts (NH_4Cl, $LiCl$, $NaCl$, KCl, $CsCl$, $MgCl_2$, $CuSO_4$ and $MnCl_2$) also increased the binding but to a lesser extant.

From the results illustrated in figure (3-8), it is suggested that these salts may provide some conformational changes in the progesterone receptors and the charged groups of the binding domain of these receptors that hinder maximal binding are shielded [172,203,204].

Among all cations studies the calcium ion was the most important for the stimulation of progesterone binding to its ovarian receptors, these results may suggest that progesterone binding is a calmoduline dependent process [205,206].

The inhibiting effect of Zn (II) ions on the progesterone binding to its receptors in these results are in agreement with those of other authors who found in there experiments that zinc was capable of binding with specific sites on steroid receptors molecule and then inhibiting the steroid binding [204,207,208,].

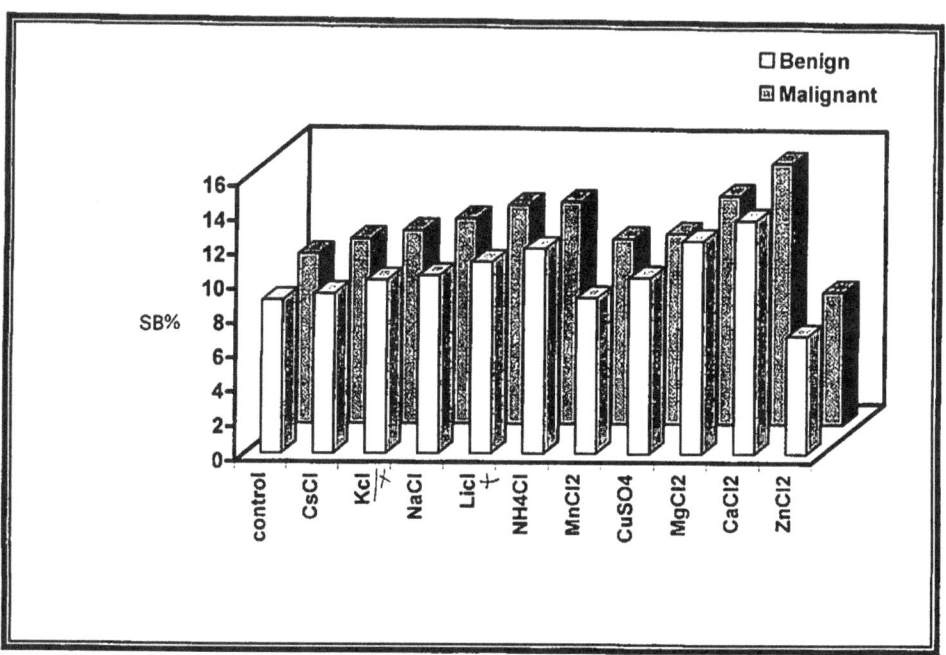

Figure (3-8): Effect of monovalent and divalent cations on the binding of ^{125}I-progesterone with its ovarian receptors.
Details are described in sections (2.5.8) and (2.5.9).

3.2.2.9 The effect of Urea on the binding of ^{125}I-progesterone with its benign and malignant ovarian tumors receptors

The effect of different concentrations of urea on the specific binding of ^{125}I-progesterone to its receptors as shown in figure (3-9). These experimental results showed that the addition of urea (concentrations ranging from 0.25 M to 4 M) resulted in a gradual dissociation of the complex. The urea at 4 M was able to dissociate about 25% of the complexes (benign and malignant), this may be attributed to the effect of urea on the hydrophobic forces participitating in the association of protein molecules [209].

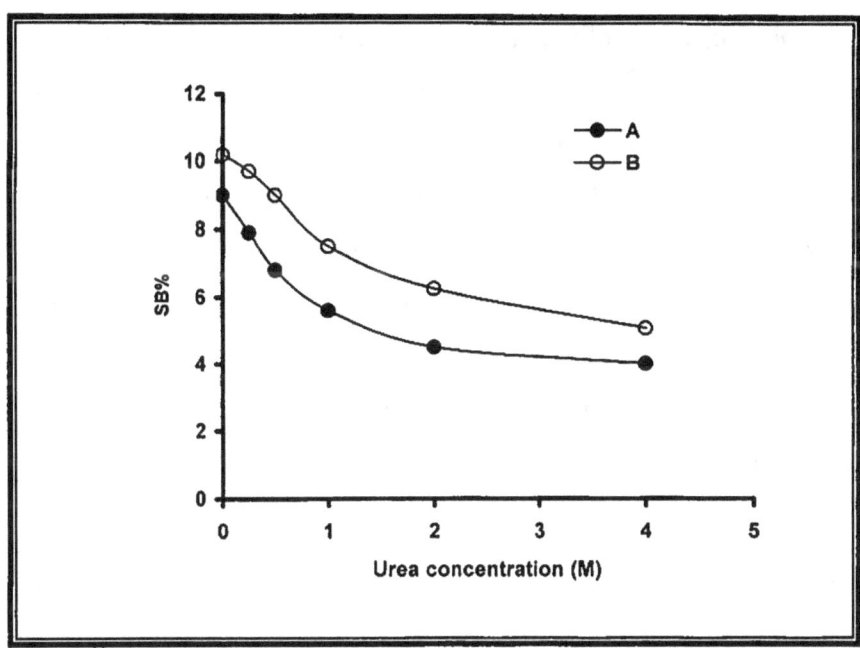

Figure (3-9): Effect of different urea concentrations on the binding of
^{125}I-progesterone with its:

A: Benign ovarian tumor homogenate.
B: Malignant ovarian tumor homogenate.
Details are described in section (2.5.10).

3.2.2.10 The effect of polyethylene glycol (PEG 10000) on the binding of ^{125}I-progesterone to its benign and malignant ovarian tumors receptors

The effect of different concentrations of PEG-10000 on the specific binding of ^{125}I-progesterone to its receptors as shown in figure (3-10). The experimental results showed that addition of PEG ranging in concentration from (2 to 10%) resulted in a gradual decrease of specific binding percent. The effect of PEG on the receptor protein solubility can be explained according to the steric exclusion mechanism proposed by Laurent [210], assuming a fixed total volume (V_T) of solvent being occupied by both polymer and protein, and defining the volume occupied by polymer as (V_E) (excluded volume i.e, volume not accessible to proteins) and the volume occupied by protein as (V'). The relation ($V_T = V' + V_E$) implies that any increase in V_E, due to increase in number or size of polymer molecules, forces a decrease in (V') and an effective increase in the concentration of protein molecules. Hence, as V_E is increased, the effective protein concentration increased as well as collision and self association of protein molecules thereby enhancing the protein precipitation because of the formation of large insoluble aggregates [173,210].

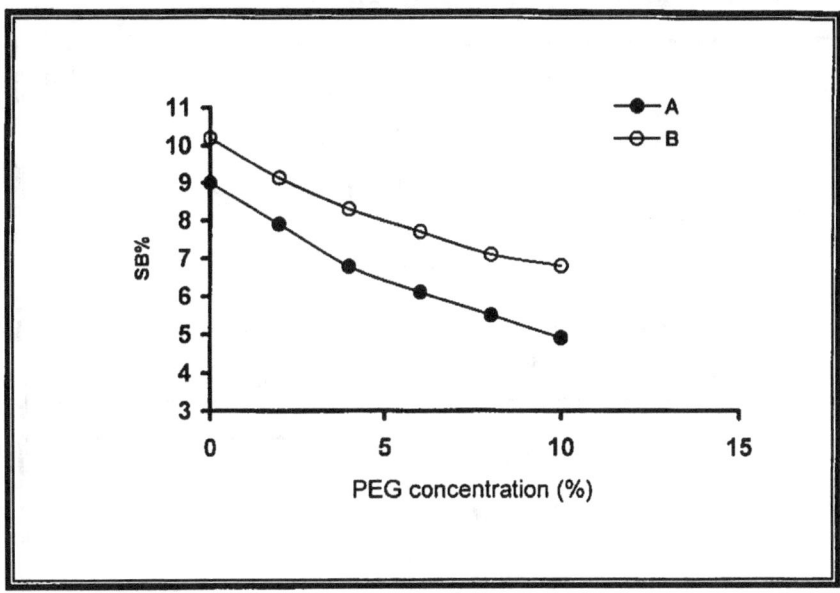

Figure (3-10): **Effect of different concentrations of PEG 10000 on specific binding of** ^{125}**I-progesterone with its:**
A: Benign ovarian tumor homogenate.
B: Malignant ovarian tumor homogenate.
Details are described in section (2.5.11)

3.3 The kinetics and thermodynamics of the interaction of progesterone with its crude cytosol receptors

3.3.1 Kinetics of the ^{125}I-progesterone binding to its crude receptors in benign and malignant ovarian tumors

The time course of the formation of ^{125}I-progesterone-receptor complex at four different temperatures (4, 25, 37 and 45 °C) in benign and malignant ovarian tumors [211,212], as shown in figure (3-11 A&B). The results of time course patterns at different temperatures revealed that the binding of ^{125}I-progesterone to its receptors in ovarian tumors is a temperature and time dependent process with a maximum binding occurs at 25°C with 6 hrs for benign receptors and at 4°C with 18 hrs for malignant receptors.

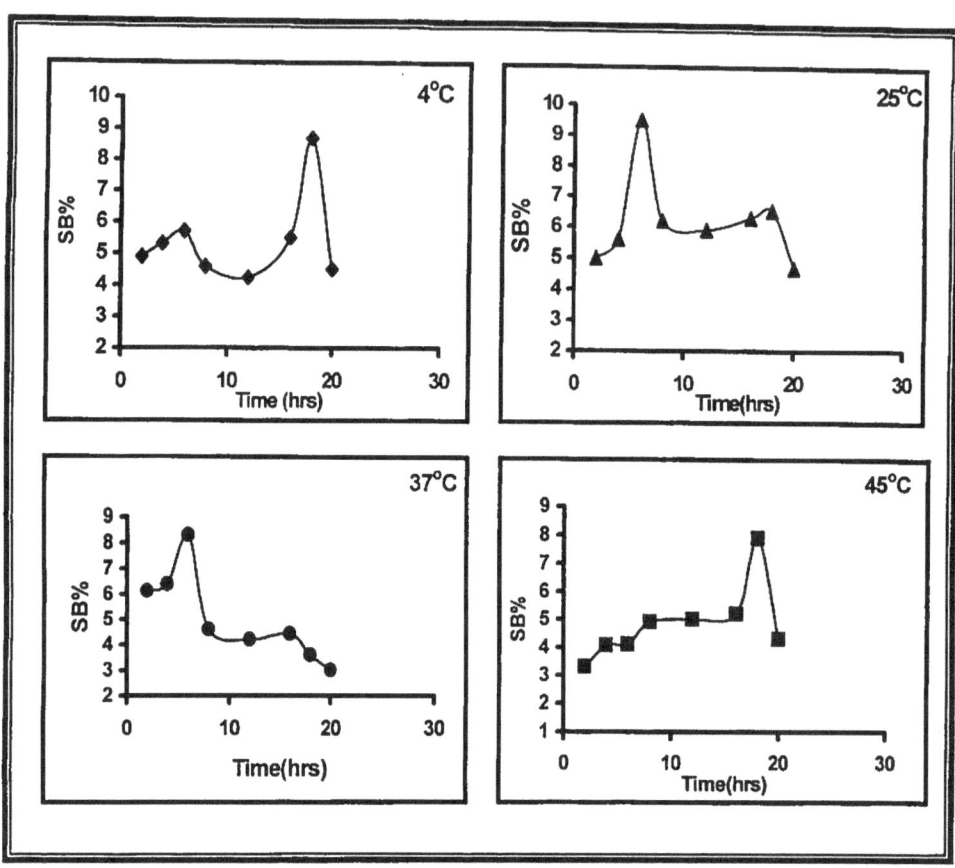

Figure (3-11): Time course of the association of[125] I-progesterone with its
A: Benign ovarian receptors at different temperatures.
Details are described in section (2.6.1).

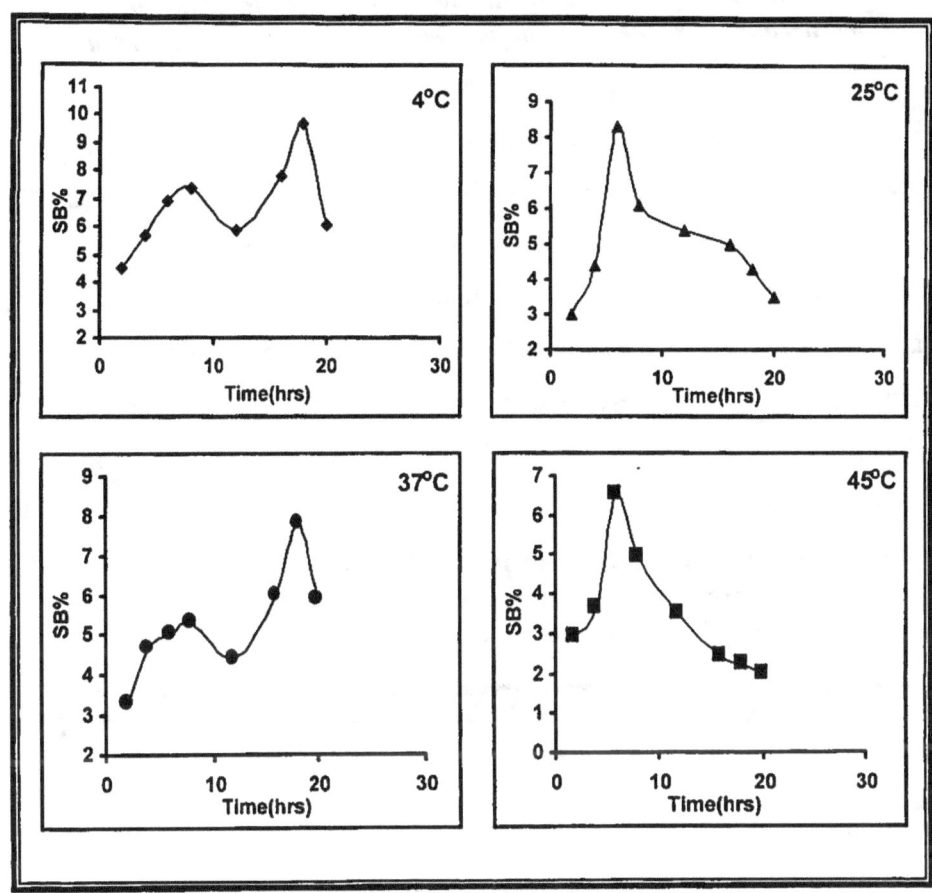

Figure (3-11): Time course of the association of[125] I-progesterone with its:
        B: Malignant ovarian receptors at different temperatures.
        Details are described in section (2.6.1).

3.3.1.1 Determination of the concentrations and affinity constants of progesterone receptors in human benign and malignant ovarian tumors

Progesterone receptors concentration and affinity constant have been measured in benign and malignant ovarian tumors. The experiment was carried out at the optimal conditions, which are obtained in previous experiments and was repeated at different temperatures (4, 25, 37 and 45°C). Scatchard plot analysis gave a straight line as shown in figure (3-12) at each temperature indicating the presence of only a single class of receptor site, or more but with the same affinity and number of binding sites. The results are summarized in table (3-2).

Table(3-2): Concentration and affinity constant of progesterone receptors in benign and malignant ovarian tumors. Details are described in section (2.6.2).

Group	Age (year)	Temp. °C	Binding capacity $B_{max} \times 10^{-9} M$	$K_a \times 10^9 M$
Premenopausal benign ovarian tumors	30 – 37	4	0.8337	0.548
		25	0.9845	0.5823
		37	0.5263	0.5149
		45	0.7015	0.2234
Premenopausal malignant ovarian tumors	29 – 35	4	1.081	0.5382
		25	0.9899	0.4184
		37	0.919	0.377
		45	0.6385	0.3548

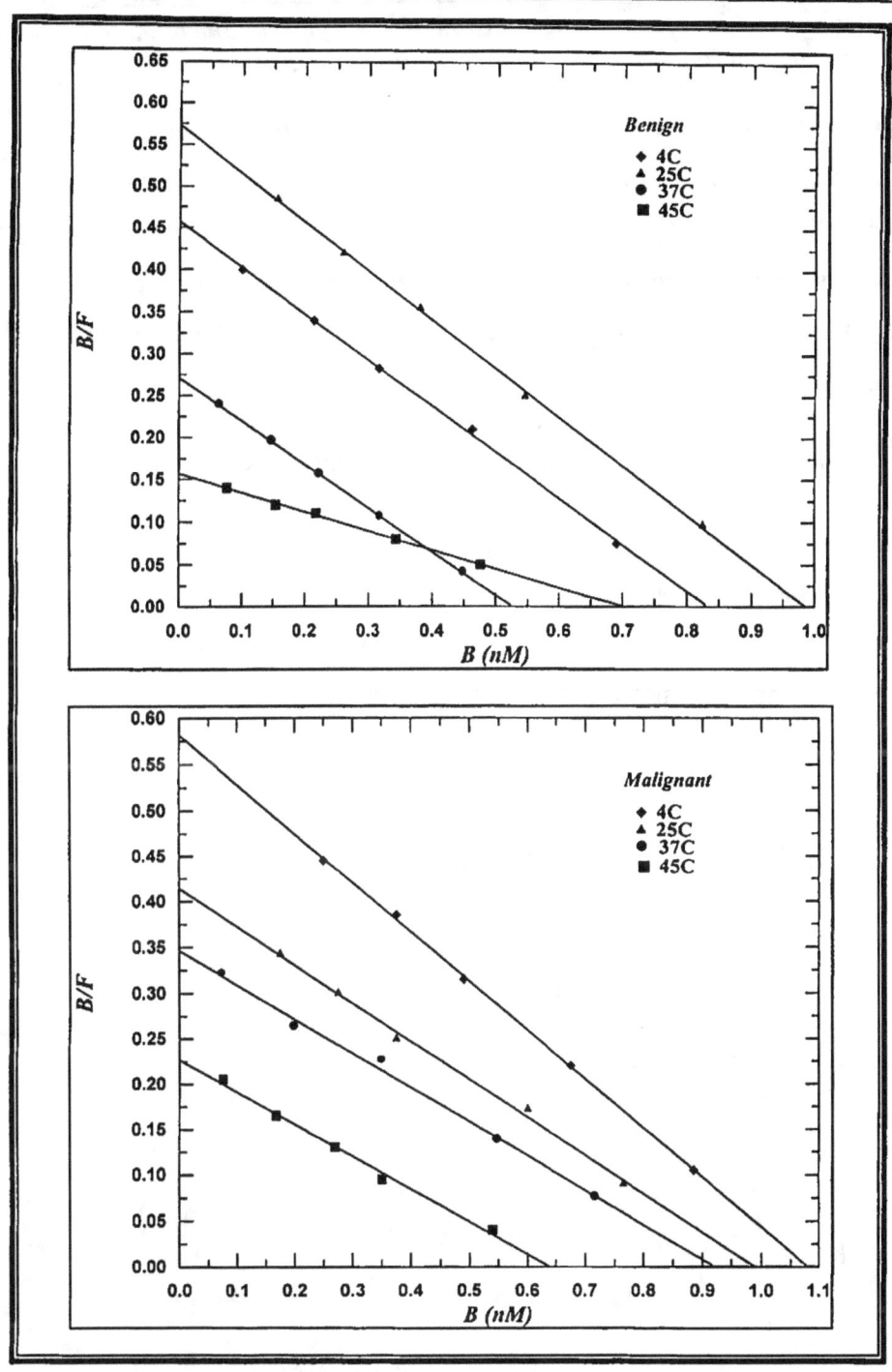

Figure (3-12): Scatchard plot of ^{125}I-progesterone binding with its receptors in
ovarian tumors homogenate at different temperatures.
Details are described in section (2.6.2).

3.3.1.2 Determination of kinetic parameters of ^{125}I-progesterone binding to its receptors in human benign and malignant ovarian tumors

The time course of ^{125}I-progesterone binding to its benign and malignant receptors in ovarian tumors was carried out to describe the kinetic parameters of the binding.

The simplest proposed model representing the interaction of ^{125}I-progesterone with its receptors could be expressed by the following equation:

$$^{125}I\text{-progesterone} + R \underset{K_{-1}}{\overset{K_{+1}}{\rightleftharpoons}} \,^{125}I\text{-progesterone - R}$$

Where:

K$_{+1}$ is the rate of the association of ^{125}I-progesterone with its receptors and K$_{-1}$ represents the rate of the reverse reaction i.e., the dissociation of the complex formed under the same conditions:

At equilibrium:

$$Ka = \frac{\left[^{125}I\text{-progesterone-R} \right]}{\left[^{125}I\text{-progesterone} \right]\left[R \right]} \quad \dots\dots\dots\dots \quad (1)$$

$$Kd = \frac{\left[^{125}I\text{-progesterone} \right]\left[R \right]}{\left[^{125}I\text{-progesterone-R} \right]} \quad \dots\dots\dots\dots \quad (2)$$

Thus ;

$$Ka = \frac{1}{Kd} = \frac{K_{+1}}{K_{-1}} \quad \dots\dots\dots\dots \quad (3)$$

Where ;

Ka is the equilibrium constant of association (affinity constant) and Kd is the equilibrium constant of the dissociation of ^{125}I-progesterone-R complex.

The values of Ka and maximal binding capacity (B_{max}) were calculated from Scatcharad at four different temperatures in figure (3-12) and table (3-2).

Table (3-3): The kinetic parameters of ^{125}I-progesterone binding to its receptors in benign and malignant ovarian tumors. Details are described in section (2.6.1) and (2.6.2).

Group	Kinetic parameters	Temperature °C			
		4	25	35	45
Premenopausal benign ovarian tumors	Binding capacity B_{max} x 10^{-9} M	0.8337	0.9845	0.5263	0.7015
	Ka x 10^{9} M^{-1}	0.548	0.5823	0.5149	0.2234
	Kd x 10^{-9} M	1.825	1.717	1.9417	4.476
Premenopausal malignant ovarian tumors	Binding capacity B_{max} x 10^{-9} M	1.081	0.9899	0.919	0.6385
	Ka x 10^{9} M^{-1}	0.5382	0.4184	0.377	0.3548
	Kd x 10^{-9} M	1.858	2.39	2.653	2.818

The kinetic association rate constant, K_{+1}, can be determined from the time course of association of ^{125}I-progesterone with its receptors and verified the order of the reaction at four different temperature. Time course data obtained from figures (3-11 A&B) could be used to confirm that the binding reaction of progesterone with its receptors in benign and malignant ovarian tumors homogenates following a first order kinetic reactions but due to the bimolecularity of this reaction, the following equation [213,214]:

$$\ln (HR)_e \left[\frac{[(H)_T - (HR)_t \, (HR)_e] / (R)_T}{(H)_T - [(HR)_e - (HR)_t}\right] = k_{+1} \, t \left[\frac{(H)_T \, (R)_T - (HR)_e}{(HR)_e} \right] \quad \text{.......(4)}$$

Equation (4) could be simplified to equation (5) when the most progesterone remained free and only a small fraction of $(H)_T$ is bound even at equilibrium (pseudo-first order) conditions:

$$\ln \frac{(HR)_e}{(HR)_e - (HR)_t} = k_{+1} \, t \left[\frac{(H)_T \, (R)_T}{(HR)_e} \right] \quad \text{...........(5)}$$

Where :

K$_{+1}$ is the kinetic association constant in (M. min)$^{-1}$.

(H)$_T$ is the total concentration of ^{125}I-progesterone.

(R)$_T$ is the total concentration of binding site.

(HR)$_e$ is the concentration of progesterone-receptor complex at equilibrium.

(HR)$_t$ is the concentration of progesterone-receptor complex at time (t).

The plotting of $\ln \dfrac{(HR)_e}{(HR)_e - (HR)_t}$ against time (t) as shown in figure (3-13), gives a straight line with a slope equal the observed value of first-order rate constant (K$_{obs}$) in (min^{-1}) and the association rate constant K$_{+1}$ was from the following formula :

$$K_{obs} = K_{+1} \frac{(H)_T (R)_T}{(HR)_e} \quad \dots \dots \dots \dots \quad (6)$$

The half life time of association (t$_{1/2}$)$_{ass}$, which represent the time needed for the formation of half amount of the complex at equilibrium, was determined from the concentration of the complex at equilibrium and the time course curve. While the half-life of dissociation (t$_{1/2}$)$_{diss.}$ was determined from :

$$(t_{1/2})_{diss.} = \frac{\ln 2}{K_{-1}} = \frac{0.693}{K_{-1}} \quad \dots \dots \dots \dots \quad (7)$$

The K$_{-1}$ value was also obtained from equation (3). Figure (3-13) represents the kinetics of complex formation between ^{125}I-progesterone and its receptors in two groups (benign ovarian tumors homogenate and malignant ovarian tumor homogenate) at different temperatures.

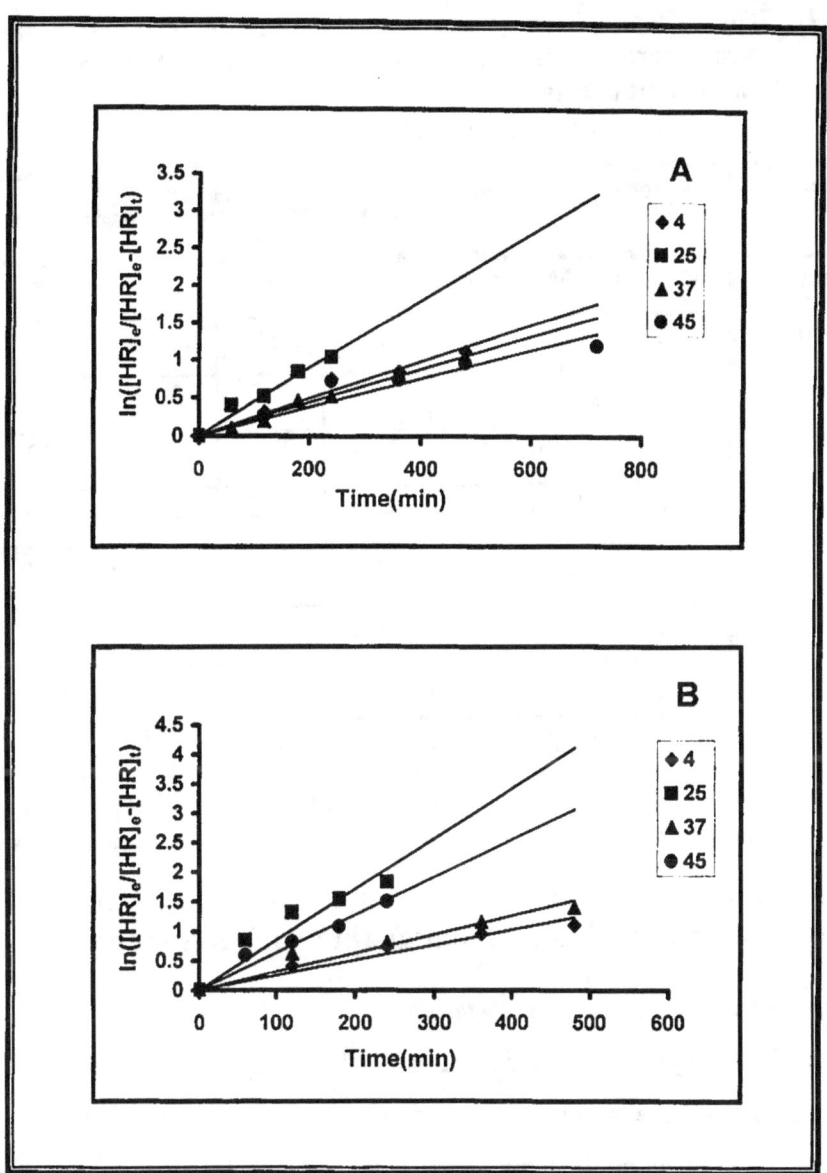

Figure (3-13): Pseudo-first order kinetics of ^{125}I-progesterone binding with its:
     A: Benign ovarian receptors.
     B: Malignant ovarian receptors.
      at different temperatures.
     Details are described in section (2.6.1)

Table (3-4): The effect of temperature on the kinetic parameter of progesterone binding to its receptors in benign and malignant ovarian tumors.
Details are described in section (2.6.1)

Group	Kinetic parameters	Temperature (°C)			
		4	25	37	45
Premenopausal benign ovarian tumors	$K_{+1} \times 10^6$ (M . min)$^{-1}$	0.258	0.439	0.378	0.283
	$K_{-1} \times 10^3$ (min)$^{-1}$	0.47	0.754	0.734	1.267
	$(t_{1/2})_{ass.}$ (hrs)	5	3	3.6	2.2
	$(t_{1/2})_{diss.}$ (hrs)	24.5	15.32	15.7	9.12
	Ka $\times 10^9$ (M)$^{-1}$	0.548	0.5823	0.5149	0.2234
Premenopausal malignant ovarian tumors	$K_{+1} \times 10^6$ (M . min)$^{-1}$	0.232	0.665	0.295	0.559
	$K_{-1} \times 10^3$ (min)$^{-1}$	0.431	1.589	0.784	1.575
	$(t_{1/2})_{ass.}$ (hrs)	5.6	2	4.5	1.5
	$(t_{1/2})_{diss.}$ (hrs)	26.8	7.27	14.7	7.33
	Ka $\times 10^9$ (M)$^{-1}$	0.5382	0.4184	0.377	0.3548

3.3.2 The thermodynamic of the binding of ^{125}I-progesterone to its receptors in benign and malignant ovarian tumors

3.3.2.1 Thermodynamic parameters of standard state

Figure (3-14) represents the dependence of the equilibrium constant (affinity constant) for the binding of ^{125}I-progesterone to its receptors in ovarian tumor homogenate on the temperature (Van't Hoff plot).

The results indicated that $\Delta H°$ in general had small values, the positive sign (for benign receptor) ascertains that the reaction were nearly endothermic while the negative value (for malignant receptor) ascertains that the reaction were nearly exothermic.

The negative values of $\Delta G°$ reflect the stability of the complex hence, the high affinity of the reactants.

The high negative values of ΔG° for the binding reactions are controlled by high positive ΔS° values as shown in table (3-5). So, our system is characterized by the sole contribution of ΔS° to the stability of the complexes formed, while ΔH° has little or no effect [215].

A high value of positive ΔS° suggests that the reaction spontaneity was entropically driven. Entropy was the driven force for the occurrence of the binding reaction. This indicates that the hydrophobic interactions played an important role in stabilizing the complex [216].

So, the negative value of ΔG° showed that the overall reaction was energetically favorable in direction of complex formation.

Table (3-5): Thermodynamic parameters at standard state of progesterone binding to its receptors in benign and malignant ovarian tumors.
Details are described in section (2.6.3)

Group	Thermodynamic parameters	Temperature (°C)			
		4	25	37	45
Premenopausal benign ovarian tumors	ΔH° (KJ/mole)	3.492	3.492	3.492	3.492
	ΔG° (KJ/mole)	-46.336	-49.99	-51.703	-50.84
	ΔS° (J/mole.K)	179.89	179.47	178.05	170.85
Premenopausal malignant ovarian tumors	ΔH° (KJ/mole)	-7.658	-7.658	-7.658	-7.658
	ΔG° (KJ/mole)	-46.29	-49.18	-50.88	-52.03
	ΔS° (J/mole.K)	139.47	139.34	139.42	139.55

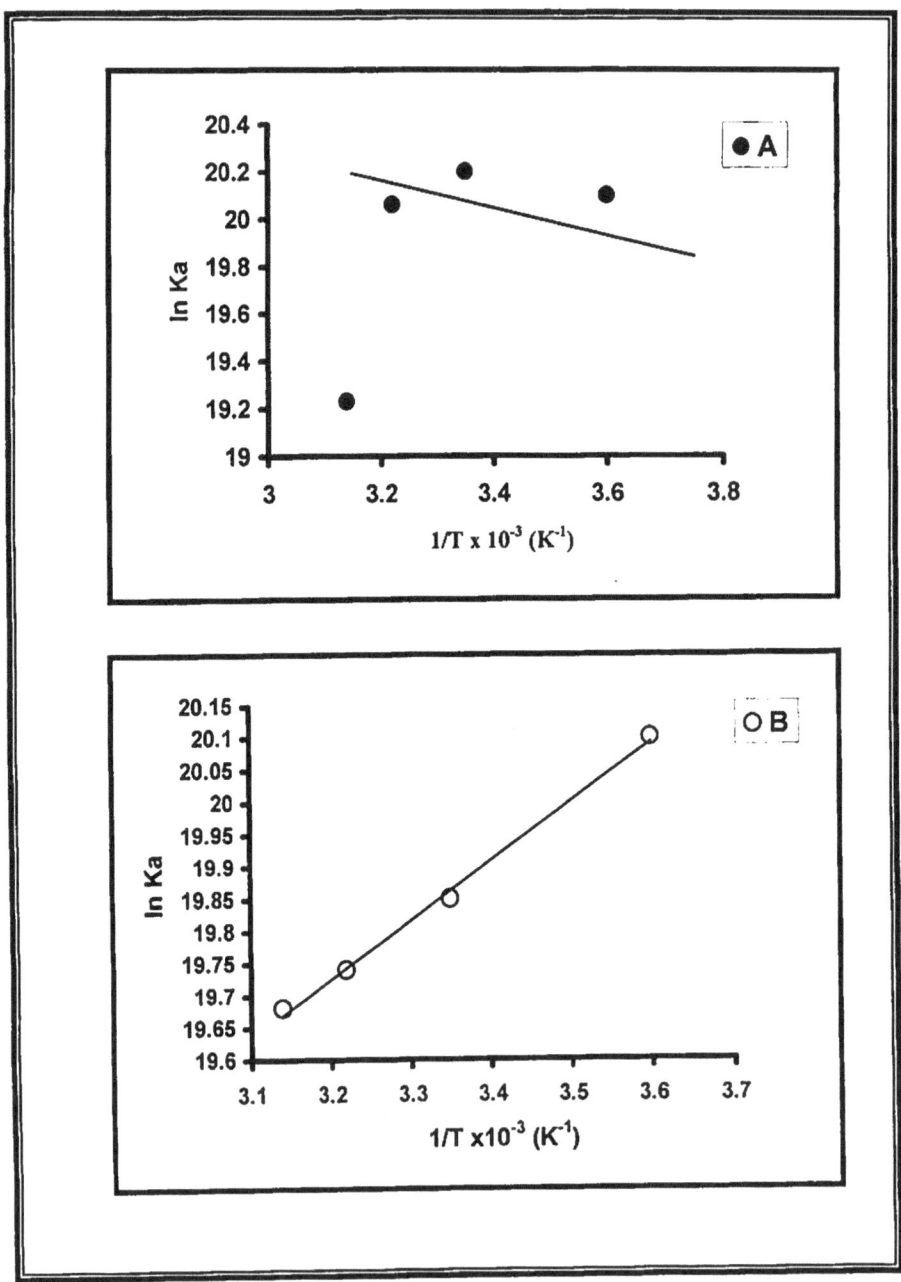

Figure (3-14): Van't Hoff plot for the ^{125}I-progesterone binding to its:
A:Benign ovarian receptors.
B:Malignant ovarian receptors.
Details are described in section (2.6.3)

3.3.2.2 Thermodynamic parameters of transition state

According to the transition state theory, the interaction of progesterone with its receptor leads to the formation of an activated complex (transition state), then the formation of the final product:

Progesterone + R ⟶ [Progesterone – R]* ⟶ Progesterone - R

An activated complex   Final product
(transition state)

The transition state thermodynamic parameters ΔH^*, ΔG^*, ΔS^* and Ea could be determined from Arrhenius equation. Figure (3-15 A&B) shows the dependence of the association rate for the binding of ^{125}I-progesterone to its receptors in ovarian tumor homogenate on temperature (Arrhenius plot).

The high positive vale of ΔG^* indicated that the formation of an activated [progesterone - R] complex was a non spontaneous process and required a lot of energy (equal to Ea) to overcome the transition state energy barrier and giving the final product, whereas the high negative ΔS^* revealed that the activated complex had a more ordered structure than the reactant species ($\Delta S^* < 0$).

The positive value of ΔG^* is mainly attributed to the decrease in entropy of the transition state ($\Delta S^* < 0$). In addition, the positive value of ΔH^* shows that the heat content of the activated complex is more than that of isolated species [217].

The results in table (3-6) show that the values of Ea and ΔH^* for the binding reactions of progesterone with its malignant ovarian receptor were more than that in the case of benign receptors. Therefore, the binding reaction of progesterone with benign receptors was easy to occur when compared with the same reaction with malignant receptors.

The negative value of ΔS^* for malignant receptors were slightly greater than that of benign receptors, so it was concluded that hydrophobic

interactions may play an important role in stabilizing the malignant receptor
activated complex formed.

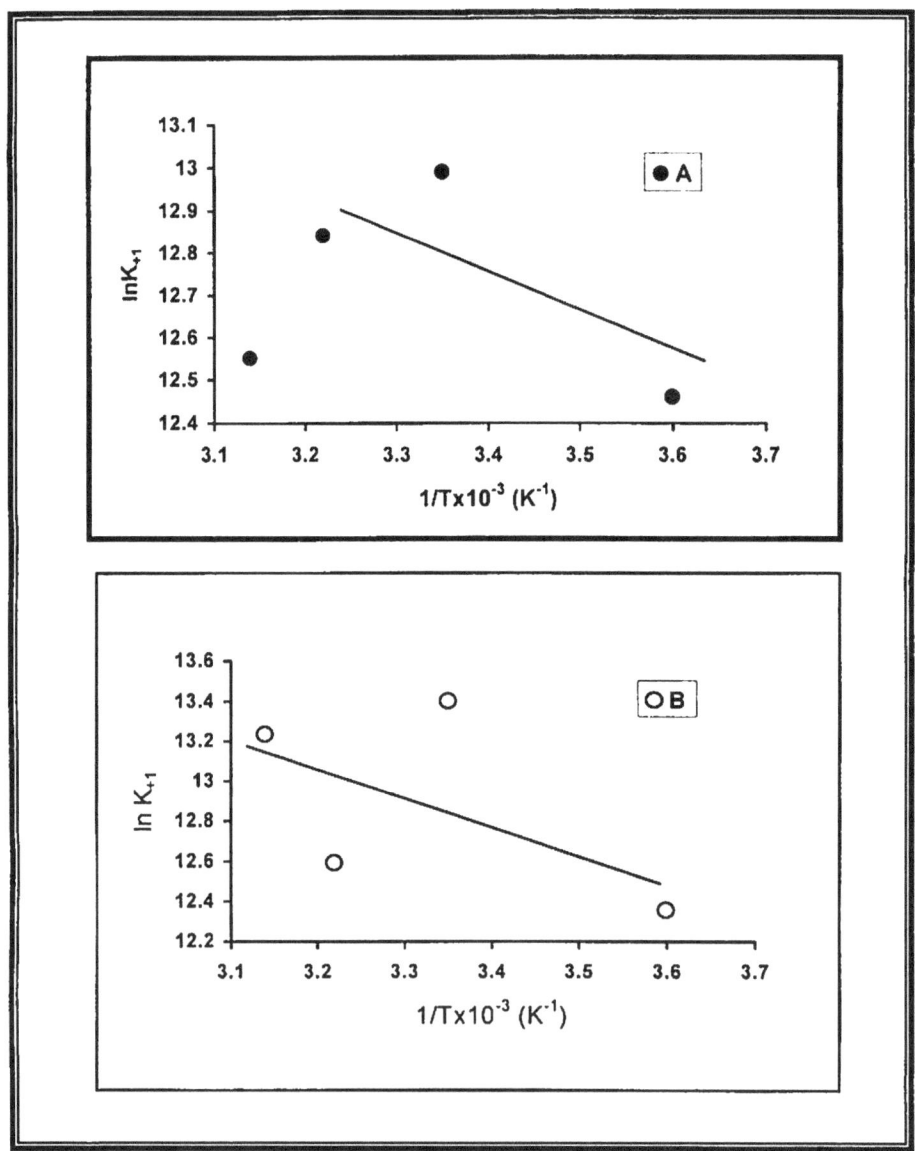

Figure (3-15): Arrhenius plot for the ^{125}I-progesterone binding to its:
        A: Benign ovarian receptors.
        B: Malignant ovarian receptors.
        Details are described in section (2.6.3)

Table (3-6): Thermodynamic parameters at transition state of progesterone binding to its receptors in benign and malignant ovarian tumors.
Details are described in section (2.6.3)

Group	Thermodynamic parameters	Temperature (°C)			
		4	25	37	45
Premenopausal benign ovarian tumors	Ea (KJ/mole)	9.312	9.312	9.312	9.312
	ΔH^* (KJ/mole)	7.009	6.834	6.73	6.67
	ΔG^* (KJ/mole)	38.978	40.80	42.93	44.87
	ΔS^* (J/mole.K)	-115.4	-113.98	-116.78	-120.15
Premenopausal malignant ovarian tumors	Ea (KJ/mole)	11.494	11.494	11.494	11.494
	ΔH^* (KJ/mole)	9.19	9.01	8.92	8.85
	ΔG^* (KJ/mole)	39.22	39.78	43.56	43.06
	ΔS^* (J/mole.K)	-108.41	-103.25	-111.76	-107.6

3.4 Isolation of cytosolic progesterone receptors using Gel Filtration Technique

Isolation of cytosol progesterone receptors were performed by gel exclusion chromatography technique. Benign and malignant homogenates were applied to Sephadex G-150 (0.7 x 23 cm) column. The void volume (V_o) of this column was (5 ml) as predicted from the elution profile of the blue dextrane. The resultant fractions of each homogenate type were collected detected for the binding with ^{125}I-progesterone as described in section (2.7.3) and (2.7.4), pooled and subjected to protein determination as mentioned section (2.4.1). This experiment revealed as shown in figure (3-16 B&C) the presence of two different eluted components (I&II), these two components eluted with different elution volume corresponding to their different molecular weights. From benign tumors homogenate, the first one (BI) eluted about (1.8) void volume (V_o) while the second one (BII) eluted with about

(3.4 V$_o$). From malignant tumors homogenate, (MI) eluted also with about (1.8 V$_o$) while the second one (MII) eluted with about (3.4 V$_o$).

Progesterone receptors have been investigated and identified in different target cells, Pichon and Milgrom [218] isolated two form^s of progesterone receptors, these two forms were sedimented on density gradient centrifugation at 8S and 4S, respectively. And also Graham et.al[219] identified two progesterone receptor protein (PR-A) M.wt (81-85 KD) and (PR-B) M.wt (116-120 KD).

Table (3-7): Purification data of progesterone receptors isolated by gel filtration technique.
Details are described in section (2.7).

Receptor type	Total protein (µg)	Specifically bound ^{125}I-progesterone (nM)	Specific binding ^{125}I-progesterone nM/mg protein	Purification factor (fold)
Crude benign ovarian tumors homogenate	250	0.599	2.396	1.00
BI isolated fraction	100	1.211	12.11	5.054
BII isolated fraction	80	0.863	10.8125	4.513
Crude malignant ovarian tumors homogenate	350	0.7476	2.136	1.00
MI isolated fraction	200	2.995	14.979	7.011
MII isolated fraction	100	1.23	12.3	5.756

BI = Benign receptor form (1)

BII = Benign receptor form (2)

MI = Malignant receptor form (1)

MII = Malignant receptor form (2)

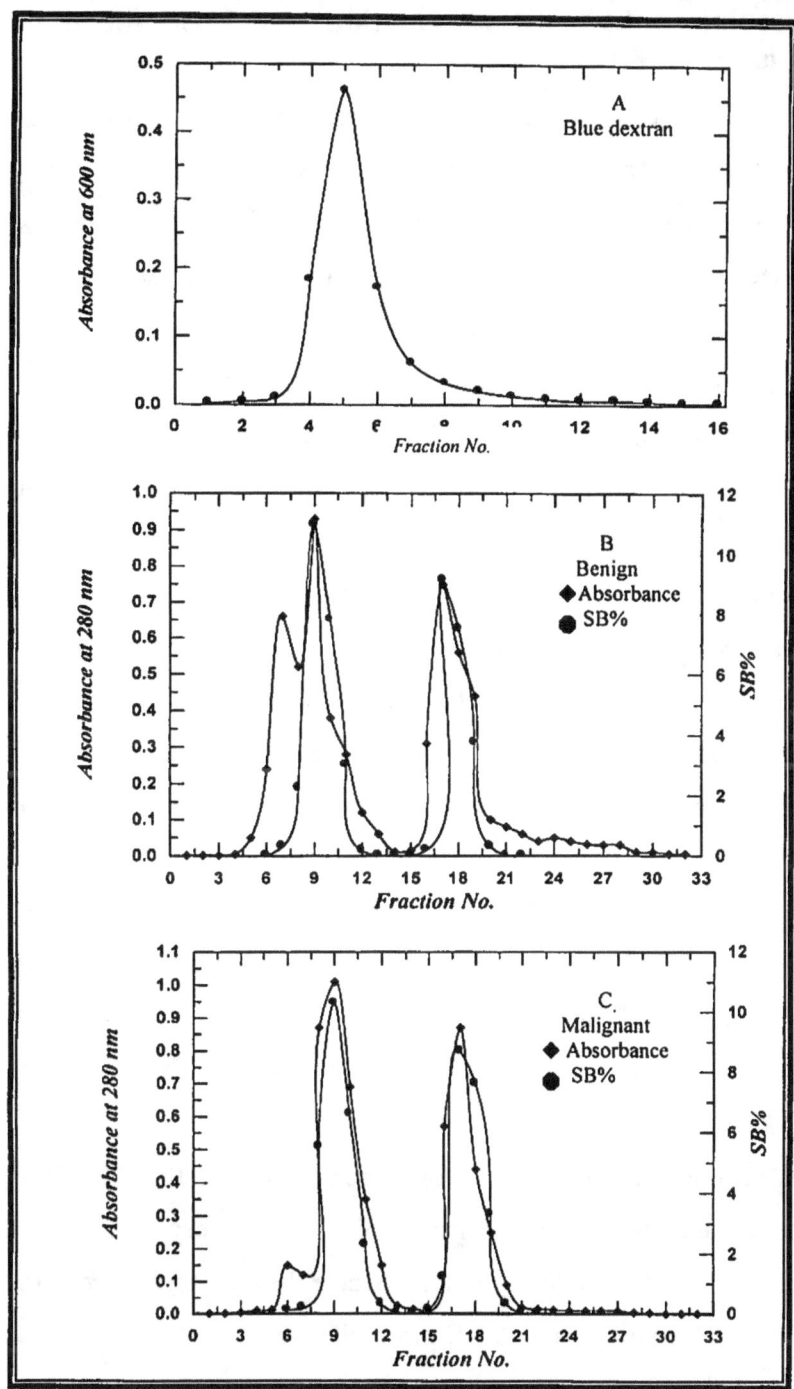

Figure (3-16): The elution profile of ^{125}I-progesterone binding with its receptors in ovarian tumors homogenate.
Details are described in section (2.7.2) and (2.7.3).

3.4.1 Molecular weight determination by gel filtration chromatography

The molecular weight of the isolated receptors was determined by gel filtration technique using sephadex gel. The void volume (V_o) figure (3-16 A), of the column was determined by using blue dextran as described in section (2.7.2) and found equal to (5 ml). Different standard proteins of known molecular weight [Ferritin (440,000), Catalase (232,000), Aldolase (158,000), BSA (67,000)] Dalton, were applied to the column and their elution volume (V_e) were measured figure (3-17). The K_{av} values for these protein were calculated by using the formula represented in section (2.7.5) and then a calibration curve was plotted between K_{av} value of the standard proteins against their logarithmic molecular weight (K_{av} vs. log $M.wt$) figure (3-18). The elution volume of the progesterone receptors were measured figure (3-17) and its K_{av} value have also been measured.

The molecular weights of progesterone receptors were estimated from the calibration curve to be (280 KD for BI and MII) and (112 KD for BII and MII).

The structures and $M.wt$ of steroid receptors isolated from different tissues were studied by many investigators. Their results have shown controversial data. Several studies revealed that steroid receptors prepared from tissues and cells not previously exposed to steroid stimulation were of $M.wt$ 250-300 KD [220].

They are two isoforms of human progesterone receptors, the 120 KD B receptor (PSR B) and the terminally truncated 94 KD A receptor (PSR A), these receptors are located in the cytoplasm, where they are associated with heat shock protein (HSP) of 70 and 90 KD molecular weight [221].

Previous study on progesterone receptors in breast tissues, revealed also two proteins by gel filtration chromatography with molecular weight 250 and 123 KD [222].

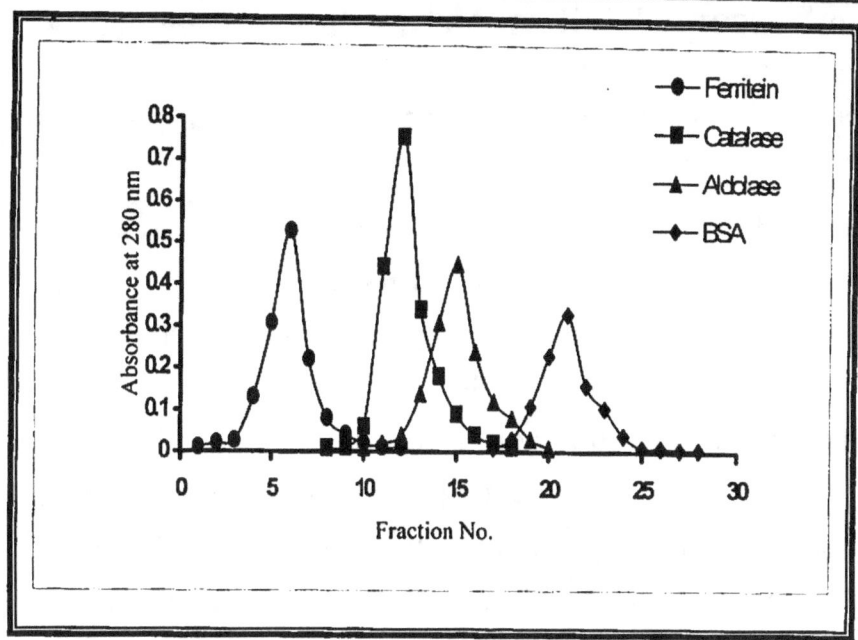

Figure (3-17): Elution profile of known molecular weight proteins.
Details are described in section (2.7.5).

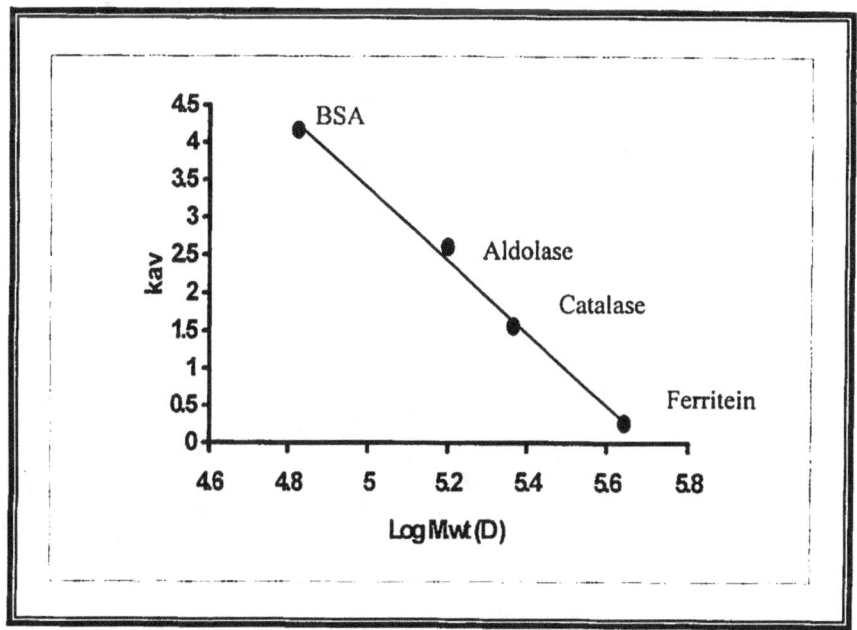

Figure (3-18): Calibration curve of known molecular weight proteins.
Details are described in section (2.7.5).

3.5 Spectroscopic studies on human progesterone receptors

3.5.1 The U.V spectra of isolated progesterone receptors in human benign and malignant ovarian tumors

Figure (3-19) illustrated that the U.V spectra of isolated progesterone receptors at pH 7.4. The U.V spectra shows that the λ_{max} for the isolated receptors BI consist of the two peaks; at 199.3 nm and 264 nm, BII-isolated receptor gives one peak at 198 nm, MI-isolated receptor gives one peak at 195 nm and MII-isolated receptors give two peaks at 196 nm and 253.3 nm.

As a result each human progesterone receptor has a characteristic spectrum and can be identified by their peaks 199.3, 198, 195 and 196 nm are assigned to tyrosine residue, while the vibrational structure as a small "wiggles" at 264 and 253.3 nm is assigned to phenylalanine [223,224].

Also it was found from figure (3-19) that tryptophan residue dose not occur on the surface of benign and malignant receptors. It seems that each of tyrosine and phenylalanine residue in the progesterone receptors in the two cases of benign and malignant is located in a way that part of it is on the surface of the protein molecule while the other part is buried.

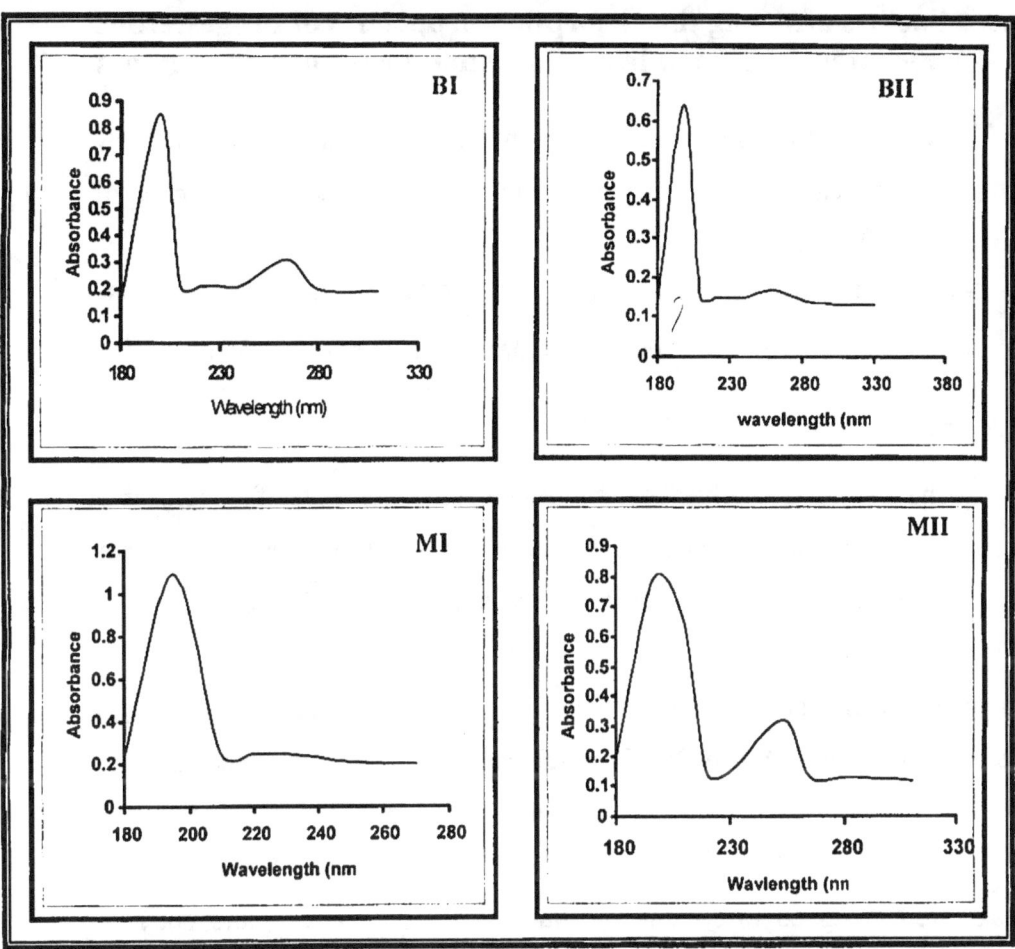

Figure (3-19): The U.V. spectra of isolated progesterone (BI, BII, MI and MII) receptors.

Details are described in section (2.8.1).

3.5.2 Factors affecting the absorption properties of progesterone receptors in human benign and malignant ovarian tumors

The absorption spectrum of a chromophore is primarily determined by the chemical structure of the molecule. However, a large number of environmental factors produce detectable changes in λ_{max} and ε. Environmental factors such as pH and polarity of the solvent provide the basis for the use of absorption spectroscopy in characterizing macromolecules [223].

• *pH effect*

The pH of the solvent determines the ionization state of the ionizable chromophore in the protein molecule. Table (3-8) shows the λ_{max} values of human progesterone receptors at different pHs (2.7, 7.4 and 10.7).

At an acidic pH 2.7, BI isolated receptors have one λ_{max} value at 203.2 nm which was assigned to phynylalanin, in BII isolated receptors three λ_{max} were obtained at 206 and 257.6 which were assigned to phenylalanine and 274 nm which was assigned to tyrosine. In MI-isolated receptors, two λ_{max} were obtained, the first one at 221 nm, the second at 276.8 nm these two peaks were assigned to tyrosine residue. In MII-isolated receptors, one λ_{max} obtained at 276 nm which was assigned to tyrosine.

When the pH value was increased from (7.4 to 10.7), an increase in the λ_{max} of tyrosine residue has been shown in all receptors type, this result is due to the dissociation of phenolic OH of tyrosine (pka = 10.07) giving an ionized from of this amino acid which absorps at higher wavelength (red shift) [223].

The spectral shift of protein produced by pH cannot be simply attributed to the inductive effects of vicinal charges, such spectral changes must therefore be attributed mainly to rearrangement of secondary and tertiary structure, although the possibility of filed effects due to unusually close conjunction of charges to aromatic groups is not excluded [224].

Table (3-8): The effect of pH on the λ_{max} of progesterone receptors spectra. Details are described in section (2.8.2.1).

pH	BI-isolated receptor	BII-isolated receptor	MI-isolated receptor	MII-isolated receptor
	λ_{max} (nm)	λ_{max} (nm)	λ_{max} (nm)	λ_{max} (nm)
2.7	203.2	206, 257.6, 274	221, 276.8	276
7.4	199.3, 264	198	195	196, 253.3
10.7	295	293.2	203, 295.5	294.1

- *Polarity effect*

The importance of this study comes from studying of the internal configuration of protein [225].

a. The effect of 20% ethanol and dimethylsulphoxide (DMSO)

Table (3-9) shows the effect of 20% ethanol and DMSO at neutral pH on the progesterone receptors spectra. In 20% ethanol, it was found that one λ_{max} was obtained for each receptor at 218, 217.5, 219.3 and 219.1 nm for BI, BII, MI and MII isolated receptors, respectively which were assigned to tryptophan residue while in the case of 20% DMSO a newer λ_{max} was appeared for each isolated receptors, these are 284.9, 288.2, 284.1 and 286 nm for BI, BII, MI and MII respectively which were assigned to tryptophan residues.

The appearance of these new λ_{max} values indicates that protein was defolded due to change in the secondary and tertiary structure of the protein that bring the tryptophan to expose to absorbance while phenylalanine and tyrosine residues were buried inside the receptor molecule, also it was found that progesterone receptors are highly sensitive to change in the polarity of the solvent.

b. The effect of 20% ethylene glycol

Table (3-9) shows the λ_{max} of human progesterone receptors BI and MI in 20% ethylene glycol at neutral pH. The data obtained previously in section (3.5.2.1) show that the λ_{max} of BI-isolated receptor at neutral pH were 199.3 and 264 nm, the λ_{max} value to tyrosine was shifted towards longer wavelength (red shift) in 20% ethylene glycol due to the hydrogen bonding of the OH groups of tyrosine with the solvent or with the π-electron system of the benzene ring where tyrosine was functioned as a hydrogen donor, while the λ_{max} value of phenylalanine was shifted towards shorter wavelengths in 20% ethylene glycol, this shift was attributed to $\pi \longrightarrow \pi^*$ transitions [223,225].

These two shifts in λ_{max} were accompanied with an increase in the absorbency of phenylanine and a decrease in the absorbency of tyrosine, these findings could be attributed to change in the protein structure that bring the phenylalanine residues to the surface of the protein while tyrosine residues were party embedded in a hydrophobic region of the protein molecule.

The changes in the protein structure for MI-receptor may bury tyrosine residues in the internal region of the protein and bring phenylalanine to the molecule surface. Also, it was found that BII and MII receptors were very high sensitive to ethylene glycol, this solvent can bury all absorbing amino acids incide the hydrophobic region of the protein molecule.

c. The effect of 20% Urea

The effect of urea on the U.V spectrum at neutral pH was examined in this experiment. Table (3-9) shows the effect of urea on the progesterone receptors U.V spectra at pH 7.4 previous data in experiment (3.5.2.1) identified the peaks in BI, BII, MI and MII-isolated receptors spectrum at pH 7.4, these peaks obtained from this experiment were assigned to tyrosine and phenylalanine. In the presence of 20% urea pH 7.4, these two amino acids were buried inside the receptor molecules and tryptophan residues were appeared on the surface of molecule with a new absorption peak, this apeared in all types of progesterone receptors BI, BII, MI and MII.

The results indicate that urea affects the progesterone isolated receptors BI, BII MI and MII structurally, since many chromophores which were embedded in an interior region of the receptor molecule where they were inaccessible to the solvent came into intact with it due to the unfolding of the molecule, and hence, different spectra were obtained [225].

Table (3-9): The effect of 20% ethanol, DMSO, ethylene glycol and urea on the λ_{max} of progesterone receptors spectra. Details are described in section (2.8.2.2).

Solvent	BI-isolated receptor	BII-isolated receptor	MI-isolated receptor	MII-isolated receptor
	λ_{max} (nm)	λ_{max} (nm)	λ_{max} (nm)	λ_{max} (nm)
20% ethanol	218	217.5	219.3	219.1
20% DMSO	284.9	288.2	284.1	286
20% Ethylene glycol	253.4	–	257	–
20% Urea	289.5	292.4	294.8	292

3.5.3 Spectrophotometric pH titration of isolated progesterone receptors in human benign and malignant ovarian tumors

Spectrophotometric pH titration is following the change in absorbance of the chromophore with increasing pH [223]. Many studies of protein structure require the determination of pka values for proton dissociation from ionizable amino acid side chains, because these values give an indication of the location of the amino acid in the protein. This can often be done spectrophotometrically because dissociation often changes the spectrum of one of the chromophores, the observation of tyrosine dissociation was performed by measuring the absorption at 259 nm (λ_{max} for the ionized from of tyrosine), and the observation of histidine dissociation was carried out by measuring the absorption at all 211 nm.

Figure (3-20 A,B) shows the pH titration curve of progesterone receptors for histidine and tyrosine respectively. (A) curves show that the pka values for histidine are 6.1, 7.2, 6.3 and 6.9 for BI, BII, MI and MII isolated receptors respectively, while the pka values for tyrosine in (B) curves were equal to 11.1, 11.5, 11.5 and 11.7 for BI, BII, MI and MII isolated receptors respectively. From the same figure it was found:

1. About 74.4, 66, 64.6 and 55% of histidine residues are located on the surface of the BI, BII, MI and MII-isolated receptors molecular respectively.

2. About 25.6, 34, 35.4 and 45% of histidine residues are embedded in the interior region of the BI, BII, MI and MII-isolated receptors omlecular respectively.

3. About 65.5, 36, 58.6 and 55.5 of tyrosine residues are located on the surface of BI, BII, MI and MII-isolated receptors molecular respectively.

4. About 34.5, 64, 41.4 and 44.5 of tyurosine residues are buried interior the folded structure of the BI, BII, MI and MII-isolated receptors molecular respectively.

The internal tyrosine residues in BII, MI and MII isolated receptors were in a strongly polar environment (e.g., a tyrosine surrounded by carboxy groups), while BI isolated receptor the tyrosine residue were largely present on the surface of the molecule and the internal tyrosine are in strongly non-polar environment. On the other hand, the histidine residues are largely present on the molecular surface of BI, BII, and MI receptors and the internal residues are in a non-polar environment whereas the internal histidine residue of MII isolated receptor are likely to be in strongly polar environment.

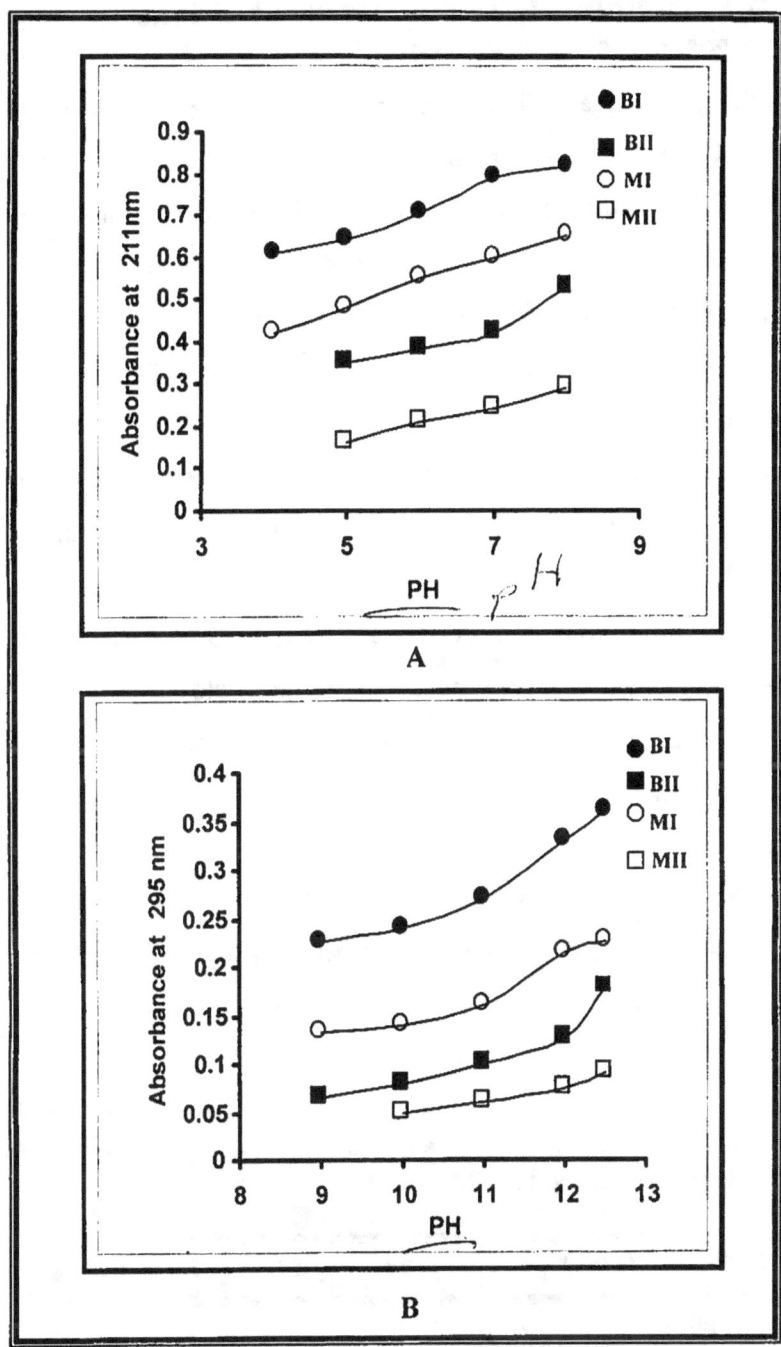

Figure (3-20): Spectrophotometric pH titration of isolation progesterone receptors,
A) histidine residues , B) tyrosine residues.
Details are described in section (2.8.3).

3.5.4 The U.V spectra of ^{125}I-progesterone and of the different ^{125}I-progesterone-receptors complexes

The binding of ligand to the active site of a receptor frequently produces spectral changes in chromophores in or near the active site by affecting the polarity of the region or the accessibility to solvent; this means that chromophores on the surface become inaccessible to the solvent by being buried in the region in which binding takes place or because a conformational change that buries or exposes a chromophore in another part of the molecule can accompany binding [223].

Table (3-10) shows the λ_{max} values of ^{125}I-progesterone and it complexes with isolated receptors. Our result indicated that the binding of progesterone with its receptor abolished the λ_{max} values of free progesterone and also tyrosine residues were embedded in an interior region of the protein and appearance of phenylalanine residues on the molecule surface in all progesterone receptor type. Our results supported the previous steroid enveloping concept which suggests that progesterone is bound by its receptors from multiple sides (α,β and peripheral) and progesterone there is being "enveloped" in the hydrophobic cavity [204].

The absorbances at 215 and 268 nm obtained for progesterone may be attributed to the n $\longrightarrow \sigma^*$, $\pi \longrightarrow \pi^*$ and n $\longrightarrow \pi^*$ electronic transition respectively.

Table (3-10): The λ_{max} values of the U.V spectra of ^{125}I-progesterone and its complexes with isolated receptors. Details are described in section (2.8.4)

Isolated receptor	λ_{max} (nm)
[125]I-progesterone-R (BI)	207
[125]I-progesterone-R (BII)	207.4
[125]I-progesterone-R (MI)	205
[125]I-progesterone-R (MII)	205.3
[125]I-progesterone	215, 268

CONCLUSIONS

1. The developed protocol for the assay of progesterone receptors is capable to analyze these receptors and the procedure is suitable for the assessment of progesterone receptors in benign and malignant serous ovarian tumors.

2. A higher incidence of progesterone receptors was obtained in malignant than in benign serous ovarian tumors in tissues.

3. The kinetic studies of the ^{125}I-progesterone binding to its receptors in premenopausal patients with benign and malignant serous ovarian tumors showed that the binding reactions is a temperature and time dependent process. The results indicate that the reaction fits pseudo-first order kinetic.

4. The results obtained from the thermodynamics studies on the association of progesterone with its receptors in benign and malignant serous ovarian tumors were spontaneously occurred ($\triangle G° < 0$) and the binding reactions were entropically driver ($\triangle S° > 0$).

5. The spectroscopic studies on progesterone receptors in premenopausal patients with benign and malignant serous ovarian tumors revealed a characteristic spectrum for each receptor.

FUTURE WORKS

1. The application of the developed method of the assessment of progesterone in serous ovarian tumors tissues to investigate the level of progesterone in other groups of patients and with other types of ovarian tumors.

2. Isolation of progesterone receptors from different types of tumors using different techniques.

3. Molecular characterization of these receptors of ovarian tumors.

4. Spectroscopic characterization of the progesterone receptors by using I.R., N.M.R, X-Ray, diffraction analysis.

References

References

1. Caret   R.L., Denniston K.J. and Topping J.J. " **Principles and Applications of Inorganic, Organic & Biological Chemistry**", 2nd ed.; The McGraw-Hill Companies. 1997, p.392.

2. Carr B.R. and Blackwell R.E. **"Textbook of Reproductive Medicine"**, 2nd ed.; Appleton & lange. 1998, p.213.

3. Guyton A.C. and Hall J.E. **"Textbook of Medical physiology"**, 9th ed.; W.B.Saunders company. 1996, p.1022.

4. Johnson L.R. **"Essential Medical Physiology"**, 2nd ed.; Lippincott-Raven. 1998, p.650.

5. Murray R.K., Granner D.K., Mayes P.A. and Rodwell V.W. **"Harper's Biochemistry"**, 25th ed; Appleton & lange. 2000, p.604.

6. Burtis C.A. and Ashwood E.R. **"Tietz Textbook of clinical chemistry"**, 3rd ed.; W.B. Saunders company, philadelphia. 1999, p.723, 1611-1612.

7. Greenspan F.S. and Gardner D.G. **"Basic and clinical Endocrinology"**, 6th ed.; The McGraw-Hill companies. 2001, p.453-458.

8. Rajan R. **"Postgraduate Reproductive Endocrinology"**, 3rd ed.; Jaypee brothers medical publishers. 1993, p.14.

9. Blick K.E. and Liles S.M. **"Principles of clinical chemistry"**; John Wiley & Sons Inc., New York. 1985, p.544.

10. Simpson E.R. **"Report of the Fourth Ross conference on obstetric Research"**; P.C. MacDonald and E.C. Hasselmeyer, Eds. 1983, p.94.

11. Carr B.R. " **Disorder of the ovary and Female Reproductive tract"**, 8th ed.; W.B.Saunders company, Philadelphia. 1992, p.733-798.

12. Adashi E.Y. **"Reproductive Endocrinology"**; W.B.Saunders company, Philadelphia. 1992, p.181.

13. Orten J.M. and Neuhaus O.W. **"Human Biochemistry"**, 10th ed.; C.V. Mosby company. 1982, p.608.

14. Daly W.J., Eston J.D., Hutton J.J., Kohler P.O., O'Rourke R.A., Sande M.A., Stein J.H., Trier J.S. and Zuaifler N.J. **"Internal Medicine"**, 2nd ed.; Little, Brown and company, Boston. 1987, p.1811-1817.

15. Kaplan L.A. and Pesce A.J. **"Clinical chemistry, theory, analysis and correlation"**, 2nd ed.; The C.V. Mosby company. 1989, p.657.

16. Fotherby K. **"Hormones in Blood"**, 3rd ed.; C.H. Gray and V.H.T. James Eds., London. 1979, p.439-491.

17. Ganong W.F. **"Review of medical physiology"**, 5th ed.; Long Medical Book. 1993, p.417.

18. Smith E.L., Hill R.L., Lehman I.R., Lefkowitz R.J, Handler P. and White A. **"Principles of Biochemistry: Mammalian Biochemistry"**, 7th ed.; The McGraw-Hill companies. 1985, p.388.

19. Davidson V.L. and Smithman D.B. **"Biochemistry"**, 4th ed. 1999, p.259-266.

20. Tortora G.J. and Grabowski S.R. **"Principles of Anatomy and Physiology"**, 9th ed.; John Wiley & Sons Inc., New York. 2000, p.570.

21. Marks D.B. **"Basic Medical Biochemistry"**; Lippincott Williams & Wilkins, A Wolters Kluwer company. 1996, p.684,742.

22. Lipsett M. and Jaffe R.B. **"Steroid Hormones"**, 2nd ed.; W.B. Saunders company, Philadelphia. 1986, p.140-153.

23. Tilton R.C., Balows A., Hohnadel D.C. and Reiss R.F.**"Clinical Laboratory Medicine"**; Mosby-Year bokk. 1992.p.285.

24. Barnes J. **"Lecture notes on Gynaecology"**, 5th ed.; Blackwell Scientific publications. 1983, p.190.

25. Speroff L., Glass R. and Kase N. **"Clinical Gynecologic Endocrinology and Infertility"**, 4th ed. 1989, p.6.

26. Seeley R.R., Stephens T.D. and Tate p. **"Essentials of Anatomy and Physiology"**, 2nd ed.; Mosby-Year book. 1996, p.532-535.

27. Treloar A.E., Boynton B.E. and Behn B.G. *Int. J. fertil.* 1967, p.12:77.

28. Pennington G.W. and Naik S. **"Hormone Analysis: Methodology and Clinical Interpretation"**, volume II; CRC press Inc. 1981, p.31.

29. Evans J.J. *Clin. Chem.* 1986, 32 (2): 542-544.

30. Ganong W.F. **"Review of Medical Physiology"**, 17th ed.; Appleton & Lang. 1995, p.408.

31. Wilson J.D., Foster D.W., Kronenberg H.M. and Larsen P.R. **"Williams Textbook of Endocrinology"**, 9th ed.; W.B. Saunders company, Philadelphia. 1998, p.761.

32. Fink G. **"The physiology of Reproduction"**, vol.1; Knobil E., Neill J.D., eds.; Raven New York. 1998, p.1349-1377.

33. Smith A.F., Beckett G.J., Walker S.W. and Rae P.W.H. **"Clinical Biochemistry"**, 6th ed.; Black well Science Ltd. 1998, p.234.

34. Edwards C.R.W., Bouchier I.A.D. and Haslett C. **"Davidson's: principles and practice of Medicine"**, 17th ed.; Churchill livingstone. 1995, p.672.

35. Pira F., Motta M. and Matini L. **"Endocrinology"**; Gruneanel Startton, New York. 1979, p.21-33.

36. Knobil E., plant T.M. and Wildt T.L. *Science.* 1980, 207: 1371.

37. Chapple S.C., Resko J.A. and Norman R.L. *J. Clin. Endocri. Metab.* 1981, 52: 1-8.

38. Wildt L., Hutchinson J.S. and Marshall G. *Endocrinology.* 1981, 109: 1293-1294.

39. Filicori M., Butler J.P. and Crowley W.F. *J. Clin. Invest.* 1984, 73: 1638-1647.

40. Liu J.H. and Yen S.S.C. *J. Clin. Endocri. Metab.* 1983, 57: 797-802.

41. Batista M.C., Cartledge T.P. and Zellmer W. *J. Clin. Endocrin. Metab.* 1992, 74: 565-570.

42. Muldoon T.G. **"Molecular Mechanism of steroid hormone action"**; Walter de Gruyter & company, Berlin. 1985, p.173.

43. Pollow K., Schmidt-Gollwitzer M. and Nerinny-Stickel J. **"Progesterone receptors in normal and neoplastic tissues"**; McGuire W.L. Raynaud J.P., Baulieu E.E., eds.; Raven, New York. 1977, p.313.

44. Pollow K. **"Hormones in normal and abnormal human tissues"**; de Gruyter W, Berlin, New York. 1981, p.378.

45. Korenman S.G. *J. Clin. Endocri. Metab.* 1968, 28: 127.

46. Hawkins R.A., Roberts M.M. and Forrest A.P.M. *Br. J. Cancer* 1980, 67: 153.

47. Lupulescu A.P. *Cancer.* 1996, 78 (11): 2264.

48. Jensen E.V. and Desombre E.R. *Science* 1973, 182: 126.

49. Kerr J.F., Winterford C.M. and Harmon B.V. *Cancer* 1994, 73: 2013.

50. Creasman W.T. and Soper J.T. *Am. J. Obstet. Gynecol.* 1985, 151: 922.

51. Bonomi P., Johnson P. and Anderson K. *Semin. Oncol.* 1985, 12: 48-54.

52. Wiest W.G. and Rao B.R. *Adv. Bio. Sci.* 1970, 7: 251.

53. Lantta M., Karkkainen J., Wahlstrom T. and Widholm O. *Acta. Obstet. Gynecol. Scand.* 1984, 63: 505-508.

54. Lantta M, *Acta. Obstet. Gynecol. Scand.* 1984, 63: 497-503.

55. Toppila M., Tyler J.P.P. and Fay R. *Br. J. obstet. Gynaecol.* 1986, 93: 986-992.

56. Agarwal N., Rao D.L. and Murgeshan K. *Int. J. Gynaecol. Obstet.* 1987, 25: 145-149.

57. Iversen O.E., Shaarland E. and Utaaker E. *Gynecol. Oncol.* 1986, 32: 65-76.

58. Tierney L.M., Mcphee S.J. and Papadakis M.A. **"Current Medical Dignosis and Treatment"**; The McGraw-Hill companies. 2001, p.746, 762.

59. Whitefield C.R. **"Dewhurst's Textbook of obstetrics and Gynaecology"**, 5th ed.; Black well science. 1995, p.759-774.

60. Devita V.T., Hellman S. and Rosenby S.A. **"Cancer principles and practice of oncology"**, 5th ed.; Lippincott-Raven. 1997, p.1502-1514.

61. Parker S.L., Tong T., Bolden S. and Wingo P.A. *Cancer Statistics*. 1996, 46: 5.

62. Cotran R.S., Kumar V. and Robbins S.L. **"Robbins pathologic Basis of disease"**, 5th ed.; W.B. Saundres company. 1994, p.1065.

63. Homer J. *Lancet*. 2000, 355: 1028.

64. Halland J.F. and Bast R.C. **"Cancer Medicine"**, 14th ed.; Williams & Wilkins. 1997, p.2295.

65. Gunderson L.L. and Tepper J.E. **"Clinical Radiation Oncology"**; churchill Livingstone. 2000, p.940-944.

66. Damjanov I. and Linder J. **"Anderson's pathology"**, 10th ed; Mosby-Year book. 1996, p.2278.

67. Fuys Woodruff J.D. **"Pathology: Practical Gynecologic Oncology"**, 2nd ed.; Williams and Wilkins company. 1995, p.1079.

68. Scully R.E. *Int. J. Gynecol. Obstet*. 1995, 49: 9-15.

69. Powell D.E., Puls L. and Van N.J. *Hum. Path*. 1992, 32: 46-47.

70. Poels L.E., Powell D.E. and De prist P.O. *Gynecol. Oncol*. 1992, 47: 53-57.

71. Jain S. and Tsai C.S. *J. Reprod. Med*. 2001, 46: 267.

72. Scott J.R., Disaia P.J., Hammond C.B. and Spellacy W.N. **"Danforth's Obstetrics and Gynecology"**, 8th ed.; Lippincott Williams & Wilkins. 1999, p.678.

73. Fausto N., Kunkei S.L., Carter H. and Carlson T. *Am. J. Pathol*. 1999, 154: 119.

74. Neijt J.P. *N. Engl. J. Med*. 1996, 334: 50.

75. Hirte H.W., Kaiser J.S. and Bacchetti S. *Int. J. Cancer*. 1994, 74: 900.

76. American Cancer Society. **"Cancer Facts and Figures"**; New York. 1992, p.11-13.

77. Partridge E.E., phillip J.L. and Menck H.R. *Cancer*. 1996, 78: 22-36.

78. "Iraq Cancer Registry Results", Ministry of Health, Iraqi Cancer Board, Iraqi Cancer Registry Center, (1986-2000).

79. Kelsey J.L. and Whittemore A.S. *Ann. Epidemiol*. 1994, 4: 89-95.

80. Daly M.B. *Hematol. Oncol. Clin. N. Am*. 1992, 6: 729-738.

81. Amos C.I. and struewing J.P. *Cancer*. 1993, 71: 566-572.

82. Greene M.H, Clark J.W. and Blayney D.W. *Semin. Oncol*. 1984, 11: 209.

83. Casagrande J.T. *Lancet* 1979, 2: 170.

84. Bland K.I., Daly J.M. and Karakousis C.P. **"Surgical Oncology"**; The McGraw-Hill companies. 2001, p.911-928.

85. Cramer D.W., Welch W.R., Cassells S. and Scully R.E. *Am. J. Obstet. Gynecol*. 1983, 147: 1-6.

86. Cramer D.W., Hutchison G.B., Welch W.R., Scully R.E. and Ryan K.J. *J. Nat. Cancer Inst*. 1983, 71: 711-716.

87. Szamborski J., Czerwinski W., Gadomska H., Kowalski M. and Wacker-Pujdak B. *Gynecol. Oncol*. 1981, 11: 8-16.

88. Hildreth N.G., Kelsely J.L., Livolsi V.A. *Am J. Epidemiol*. 1981, 114: 398-405.

89. Lee N.C., Wingo P.A., Gwinn M.L. *N. Engl. J. Med*. 1987, 316: 650-655.

90. Cramer D.W., Hutchison G.B., Welch W.R., Scully R.E. and Knapp R.C. *N.Engl. J. Med*. 1982, 307: 1047-1051.

91. Cetal B. *Int. J. Gynecol. Pathol*. 1992, 11: 180.

92. Lynch H.T. and Lynch P.M. *Am. J. Surg*. 1979, 138: 439.

93. Lynch H.T., Watson P. and Bewtra T.A. **Cancer** 1991, 61: 1460.

94. Frank T.S. *Cancer Control* 1999, 6: 327.

95. Stratlon J.F., Gayther S.A. and Russel P. *N. Engl. J. Med*. 1997, 336: 1125.

96. Peckham M., Pinedo H., Veronesi U. **"Oxford Textbook of Oncology"**; Oxford University press Inc. 1995, P.1294-1298, 1352.

97. Porth C.M. **"Pathophysiology"**, 4th ed.; J.B. Lippincott company, Philadelphia. 1994, p.769-770.

98. Benson R.C. and pernoll M.L. **"Handbook of obstetrics and Gynecology"**, 9th ed.; LoaAtlos, CA: Lange Madical Publications. 1994, p.563.

99. Chamberlain G. **"Gynaecology By Ten Teachers"**, 16th ed.; Edward Arnold company. 1995, p.142-146.

100. Ryan K.J., Berkowitz R. and Barbieri R.L. **"Kistner's Gynecology: Principles and practice"**, 5th ed.; 1990, p.276.

101. Hart D.M., Norman J., Callander R. and Ramsden I. **"Gynaecology illustrated"**, 5th ed.;Churchill Livingstone. 2000, p.265-267.

102. Beckman C.R.B. and Ling F.W. **"Obstetrics and Gynecology"**, 2nd ed.; 1996, p.460.

103. Roulston J.E., Leonard R.F. and Bagshawe K.D. **"Serological Tumar Markers"**; Langman Singapora. 1993, p.60-61.

104. Wharton J.T. and Herson J. *Cancer*. 1981, 48: 582-589.

105. Hacker N.F., Berek J.S., Lagasse L.D., Nieberg R.K. and Elashaff R.M. *Obstet. Gynecol*. 1983, 61: 413-420.

106. Danplat J., Ferriere J.P. and Gorbinet M. *Cancer*. 1986, 57: 1627-1631.

107. Gershenson D.M., Copeland L.J. and Wharton J.T. *Cancer*. 1985, 55: 1129-1135.

108. Hudson C.N. and Chir M. *Gynecol. Oncol*. 1973, 1: 370-378.

109. Howkins J. and Bourne G. "Shaw's Textbook of Gynaecology", 9th ed., B.I. publications PVT. LTD. 1976, p.658.

110. Pirer M.S. *Cancer*. 1984, 54: 2706-2715.

111. Richardson G.S., Scully R.E., Nikrui N. and Nelson J.H. *N. Engl. J. Med*. 1985, 312: 415-424 and 474-483.

112. Rosenshein N.B., Leichner P.K. and Vogelsang G. *Obstet. Gynecol. Surv*. 1979, 34 (suppl): 708-720.

113. Steiner M., Rubinov R., Borovik R., Cohen Y. and Robinson E. *Cancer*. 1985, 55: 2748-2752.

114. Casciato D.A. and Lowitz B.B. *"Manual of Clinical Oncology"*, 4th ed., Lippincott Williams and Wilkins. 2000, p.259.

115. Covens A., Bouchers S., Roche K., Macdonald M., Pettitt D., Jolain B., Souetre E. and Rivierc M. *Cancer*. 1996, 77: 2086.

116. Kohler D.R. and Gdldspiel B.R. *Pharmaco Therapy*. 1994, 14:3.

117. Ozola R.F. *Ann. Med*. 1995,27: 127.

118. Ward H.W.C. *Br. J. Obstet. Gynaecol*. 1972, 79: 555-559.

119. Timothy I. *Br. J. Obstet. Gynaecol*. 1982, 89: 561-563.

120. Jolles B. *Br. J. Cancer*. 1962, 16:209-221.

121. Backstrom T., Mahlck C.G. and Kjellgren O. *Gynecol. Oncol*. 1983, 16: 129-138.

122. Jolles C.J., Freedman R.S. and Jones L.A. *Gynecol. Oncol*. 1983, 16: 352-359.

123. Myers A.M., Moore G.E. and Major F.J. *Cancer*. 1981, 48: 2368-2370.

124. Holt J.A., Caputo T.A., Kelly K.M., Green Wald P. and Chorost S. *Obstet Gynecol*. 1979, 53: 50-58.

125. Jame O., Kauppila A., Syrjala P. and Vihka R. *Int. J. Cancer*. 1980, 25:175.

126. Kuhneil R., De Graaf J., Rao B.R. and Stolk J.G. *J. Steroid Biochem.* 1987, 26: 393.

127. Rendina G.M., Dodadio C. and Giovannini M. *Eur. J. Gynaecol. Oncol.* 1982, 3: 241-246.

128. Vierikko P., Kanppila A. and Vihko R., *Int. J. Cancer.* 1983, 32: 413-422.

129. Willcocks D., Toppila M., Hudson C.N., Tyler J.J.P., Baird P.J. and Estman C.J. *Gynecol. Oncol.* 1983, 16: 246-253.

130. Sutlonn G.P, Senior M.B., Strauss J.F., Mikuta J.J. *Gynecol. Oncol.* 1986, 23: 176-182.

131. Quinn M.A., Pearce P., Rome R., Funder J.W., Fortune D. and Pepperell R.J. *Br. J. Obstet Gynaecol.* 1982, 89: 754-759.

132. Bradburg J. *Lancet.* 2000, 356: 1826.

133. Kennedy C.R. and Gordon H. *Br. J. Obstet. Gynaecol.* 1981, 88: 1186-1191.

134. Yancik R. *Cancer.* 1993, 74: 1995.

135. Salls and Stone M.L. *Prog. Clin. Cancer.* 1973, 5:249.

136. Clark-Pearson D.L. and Dawood M.Y. **"Green's Gynecology: Essentials of clinical practice"**, 4th ed.; Little, Brown and company. 1990, p.531-541.

137. Slotman B.J. and Rao B.R. *Anticancer research.* 1988, 8: 417-434.

138. Pandha H.S., Waxman J. and Sikora K. *Br. J. Hospital Medicine.* 1994, 51(6): 297.

139. Van Nagell J.R., Donaldson E.S., Hauson M.B., Gay E.C. and Pavlik E.J. *Cancer.* 1981, 48: 495.

140. Wild D. **"The Immunoassay Handbook"**, 2nd ed.; Nature publishing group, U.K. 2001, p.639.

141. Pamies R.J. and Crawford D.R. *Med. Clin. N.Am.*1996, 80: 185-199.

142. Bast R.C. and Knapp R.C. *Semin. Oncol.* 1984, 11: 264-274.

143. Kenemans p., Yedema C.A. and Hilgers J.H.M. *Eur. J. Obstet. Gynecol. Reprod. Biol.* 1988, 29: 207-218.

144. Fritche H.A. and Bast R.C. *Clin. Chem.* 1998, 4 (7): 1379-1380.

145. Blaakear J., Hogdall C.K., Micic S., Toftager L.K., Hording U., Bennett P. and Bock J. *Eur. J. Obstet. Gynecol. Repord. Biol.* 1995, 59 (1): 53-56.

146. Bast R.C., Klug T.L. and John S.T. *N. Engl. J. Med.* 1983, 309: 883-887.

147. Zurawski V.R., Orjaseter H. and Andersen A. *Int. J. Cancer.* 1988, 42: 677-680.

148. Schilthuis M.S., Alders J.G. and Bouma J. *Br. J. Obstet. Gynaecol.* 1987, 94: 202-207.

149. Briosch, P.A., Irion O., Bischof P., Boder M., Forni M. and Krauer F. *Br. J. Obstet .Gynaecol.* 1987, 94: 196-201.

150. Altaras M.M., Goldberg G.I., Levin W., Bloch B., Darge I. and Smith J.A. *Gynecol. Oncol.* 1988, 30: 26.

151. Alrarez R.D., Boots L.R., Shingleton H.M., Hatch K.D., Hubbard J., Soong S.j. and Potter M.E. *Gynecol. Oncol.* 1987, 26: 284.

152. Gold P. and Freedman S.O. *J.Exp. Med.* 1965, 122: 439.

153. Malkin A., Kellen. J.A., Lickrish G.M. and Bush R.S. *Cancer.* 1978, 42: 1452-1456.

154. Van Nagell J.R., Donaldson E.S., Gay E.C., Sharkey R.M., Rayburn P. and Goldenberg D.M. *Cancer.* 1978, 41: 2335-2340.

155. Khoo S.K., Warner N.L., Lie J.T. and Mackay I.R. *Int. J. Cancer.* 1973, 11: 681.

156. Crum G.P. and Fenoglio C.M. *Diagn. Gynecol. Oncol.* 1980, 2 (3): 103.

157. Ludwing H. *Eur. J. Obstet. Gynacol. Reprod. Biol.* 1988, 28: 104.

158. Van Nagell J.R., Kim E., Casper S. *Cancer Res.* 1980, 40: 502-506.

159. Vaitukaitis J.L. and Braunstein G.P. *Am. J. Obstet. Gynecol.* 1972, 113: 751.

160. Bajshar W.K.D. *Cancer.* 1976, 38: 1373.

161. Rayter Z. *Br. J. Surg.* 1991, 78: 528.

162. Saarikoski S., Selander K., Kallio S. and Pystynen P. *Gynecol. Obstet. Invest.* 1982, 13: 206-212.

163. Friedlander M.L., Quinn M.A. and Fortune D. *Gynecol. Oncol.* 1989, 32: 184-190.

164. Toppila M., Tyler J.P.P., Fay R., Baird P.J., Crandom A.J., Eastman C.J. and Hudson C.N. *Br. J. Obstet. Gynaecol.* 1986,93: 986-992.

165. Schwartz P.E., Merino M.J., Lirolsi V.A., Lawrence R., Maclusky N. and Eisenfeld A. *Obstet. Gynecol.* 1985, 66: 428-433.

166. Slotman B.J., Kuhnel R., Rao B.R., Dijkhuizen G.H., Graaff J. and Stolk J.G. *Gynecol. Oncol.* 1989,33: 76-81.

167. Ford L.C., Berek J.S. and Lagasse L.D. *Gynecol. Oncol.* 1983, 15: 299-304.

168. Bizzi A., Codegoni A.M. and Landoni F. *Cancer Res.* 1988, 48: 6222-6226.

169. Kauppila A., Vierriko P., Kivinen S., Stenback F. and Vihko R. *Obstet. Gynecol.* 1983, 61: 320-326.

170. Munstedt K., Steen J., Knauf A.G., Buck T., Georgi R.V. and Franke F.E. *Cancer.* 2000, 89 (8): 1783-1791.

171. Lee B.H., Hecht J.L., Pinkus J.L. and Pinkus G.S. *Am. J. Clin. Pathol.* 2002, 117 (5): 745-750. [Abs].

172. Green B. and Leake R.E. "Steroid Hormone A practical Approach"; IRL press. Limited. 1987, p. 64-82.

173. Deutscher M.P. "Methods in Enzymology", volume 182; Academic press Inc. 1990, 197-306.

174. (Clinical assays, Gamma coat progesterone-^{125}I- RIA Kit); Immunotech-France.

175. Morris B.J. *Clinica. Chem. Acta.* 1979, 73: 213.

176. Lowry O.H., Roscbrough N.J., Farr A.L. and Randall R.J. *J.Biol. Chem.* 1951, 193: 265-275.

177. Slotman B.J. and Rao B.R. *The Cancer Journal* 1989, 2 (11): 373-377.

178. Scatchard G. *Ann. Ny. Acad. Sci.* 1949, 51: 660.

179. Segel I.H. **"Biochemical Calculations"**, 2nd ed.; John Wiley & Sons., Inc. 1976, p.241, 200, 278-281.

180. Scopes P.K. **"Protein Purification: Principles and practice"**, 2nd ed.; Springer-Verlage. 1987, p.196-198.

181. Gel filtration Leaflet, theory and practice; Pharmacia-Fine chemicals, p.46.

182. Farrant T.J. **"Practical Statistics for the Analytical Scientist"**; LGC. 1997, p.16, 49.

183. Euan H.D. *Clin. Chem.* 1973, 19 (12): 1403-1408.

184. Haynes S.P., Corcoran J.M., Estman C.J. and Doy F.A. *Clin. Chem.* 1980, 26 (11): 1607-1609.

185. Blight L.F. and White G.H. *Clin. Chem.* 1983, 29 (6): 1024-1027.

186. Heinonen P.K., Koivula T. and Pystynen P. *Acta. Obstet. Gynecol. Scand.* 1985, 64: 649-652.

187. Heinonem P.K., Morsky P., Aine R., Koivula T. and Pystynen P. *Maturitas.* 1988, 9: 325-338.

188. Marinaccio M., Putignano G., Geuse S., Quaranta M., Schonauer L.M., Latiano T., Stanziano A., Alfonso R. and Del Bianco A. *Eur. J. Gynaecol. Oncol.* 2000, 21 (4): 423-425. [Abs.].

189. Berqqvist A., Kullander S. and Thorell J. *Acta. Obstet. Gynecol. Scand. (suppl.)* 1981, 101: 75-81.

190.Creasman W.T., Sasso R.A., Weed J.C. and McCarty K.S. *Gynecol. Onocol.* 1981, 12: 319-327.

191.Jones L.A., Edwards C.L., Freedman R.S., Tan M.T. and Gallager H.S. *Int. J. Cancer.* 1983, 32: 567-571.

192.Brinkinshaw M. and Falconer I.R. *J. Endocrinol.* 1977, 55: 323.

193.Shiu P.R.C. and Friesen H.G. *Biochem. J.* 1974, 140: 310.

194.Fridvich I. *Ann. Rev. Biochem.* 1975, 44: 147.

195.Haro L.S. and Talamantes F.G. *Molec. Cell. Endocri.* 1985, 43: 199-204.

196.Zaltsman Y.A. and Saloman Y. *Endocrinology* 1980, 106: 1166.

197.Raganiemi H.J., Rounberg L., Kauppila A. and Ylostalo P. *J. Clin. Endocrinol. Metab.* 1981, 108: 307.

198.Al-Omar B.S. (1983), **"Molecular Characterization of Progesterone Receptors in Uterine Endometrium Effected by Tumor"**, M.Sc. thesis, supervised by Al-Mudhaffar S.A., College of Science, Baghdad University.

199.Daxembichler G., Grill H.J., Wiesinger H., Wittliff J.L., Dapunt O. **"Multiple Molecular Forms of Steroid Hormone Receptors"**; Elisevier, North Holland Biomedical press. 1977, p.163.

200.Rao M.C., Richards J.S., Midgley A.R., Leo J.R. and Reichert E. *Endocrinology.* 1977, 101: 512.

201.Lee C.Y. and Ryan R.J. *Biochemistry.* 1973, 12: 4609.

202.Melander W.and Horrath C. *Arch. Biochem. Biophysi.* 1977, 183: 200-215.

203.Walsh M.P., Vallet B. and Autric F. *J. Biol. Chem.* 1979, 254: 12136-12144.

204.Litwack G. **"Biochemical Actions of Hormones"**, volume IV; Academic press. 1977, p.358, 364, 371-372, 394.

205. Joshi L.R., Borland S.R., Hewlett E.L. and Katz M.S. *Arch. Biochem. Biophys.* 1988, 261: 134.

206. Gong Y., Block L.J. and Perry J.E. *Endocri.* 1995, 163 (5): 2172-2178.

207. Joan R.M. and Stitch S.R. *J. Endocrino.* 1973, 58: 405-419.

208. Leak A., Chrisholm G.D., Busuttil A. and Habib F.k. *Acta. Endocri. Copenh.* 1984, 105: 281-288.

209. Haro L.S. and Talamante F.G. *Mol. Cell. Endocri.* 1985, 41: 93.

210. Laurent T.C. *Biochem. J.* 1963, 89: 249.

211. Hsuch A. J. W., Peak E.J. and Clark J. H. *Steroids.* 1974, 24: 599-611.

212. O'Malley B.W. and Birnbaanmer L. **"Receptors and Hormone Action"**, volume I; Academic press. 1977, p.396-406.

213. Seely D.H., Wang W.Y. and Salhanick H.A. *Biochem. Biophys. Acta.* 1980, 632: 536.

214. Weiland G.A. and Molinoff P.B. *Life Science* 1981, 29: 314.

215. Nemethy G. and Scherage H.A. *J. Phys. Chem.* 1962, 66: 1773.

216. Waelbroeck M., Van obberghen. and Demeyts P. *J. Biol. Chem.* 1979, 254: 7736.

217. Laporte D.C., Wierman B.M. and Storm D.R. *Biochem.* 1980, 19: 3814.

218. Pichon M.F. and Milgrom E. *Cancer Res.* 1977, 37: 464.

219. Graham J.D., Yeates C. and Balleine R.L. *Cancer Res.* 1995, 55: 5063.

220. Sherman M.R. and Sterens J. *Ann. Rev. Physiol.* 1984, 46: 83.

221. Kadhum M. (1996), **"Evaluation of Some Biochemical Constituents (Enzymes and Trace Elements) in Breast Tumor Patients"**, ph.D. thesis, supervised by Al-Mudhaffer S.A., College of Science, Baghdad University.

222. Stefaneaun L., Sasano H. and Kovacs K. **"Molecular and Cellular endocrine pathology"**; Arnold. 2000, p.433.

223. Freifelder D. **"Physical Biochemisty"**, 2nd ed.; W.H. Freeman and company. Sanfrancisco. 1982, p.500-503, 511-517.

224. Yanari S. and Borey F.A. *J. Bio. Chem.* 1960, 235 (10): 2818-2825.

225. Leach S.J. and Scheraga H.A. *J. Biol. Chem.* 1960, 235 (10): 2827-2829.